Macmillan Geography

General editors:
Keith Orrell and Harry Tolley

**This book is to be returned on or before
the last date stamped below.**

M
MACMILLAN
EDUCATION

First published 1984
Reprinted 1986, 1987

Published by
MACMILLAN EDUCATION LTD
Houndmills, Basingstoke, Hampshire RG21 2XS
and London
Companies and representatives
throughout the world

Printed in Hong Kong

British Library Cataloguing in Publication Data
Hilton, Keith
Understanding Landforms. — (Macmillan Geography)
1. Physical geography — Textbooks — 1945-
I. Title
910'.02   GB55
ISBN 0-333-27643-4

Preface

In recent years rapid changes in Geography have left teachers with a wide, almost bewildering choice of content and methodology for inclusion in their courses. Thus in devising this series and advising on each text we, the editors, have had regard for the needs of pupils and the general educational aims of schooling as well as the evolving nature of the subject.

The series is intended primarily for 14-16 year olds studying for GCSE examinations, though it may be used profitably with students following new sixth form courses. As editors, we have tried to include the best of what Geography has to offer the school curriculum at this level. Specifically, we think the subject can contribute to a greater understanding of the pupils' real world experience, their acquisition of a sense of place, the development in them of an environmental ethic and an empathy with people from different cultures and in other societies. Additionally, Geography can aid the development of literacy, numeracy and graphicacy as well as intellectual and social skills. These subject specific and general goals have been used as criteria in the writing and editing of this series.

The series presents a diversity of resources on which exercises have been based to demonstrate the principles of enquiry in Geography and to promote general study skills and competence. We have tried to match the reading level of the text to the abilities of 14–16 age groups. An appropriate geographical vocabulary has been deliberately introduced and defined in each of the texts.

The topics which are covered within the five titles in the series are the ones which, in the opinion of the authors and editors, have greatest relevance to the future lives of pupils. We have tried to demonstrate how the geographer's view can contribute to an understanding of issues and problems at a variety of scales in a diversity of locations.

When looking at textbooks for the 14–16 age group the impression is often given that many of the interesting new

Contents

developments have taken place in human geography and that physical geography has lagged far behind. In this book Keith Hilton amply demonstrates that this is far from being the case. Modern ideas on the physical and human processes which shape the earth are dealt with in an informed, interesting and imaginative way. The text is well illustrated with a large number and wide range of stimulating examples and resources. A rich variety of student exercises and activities adds to the value and attractions of the book from the teacher's point of view.

Acknowledgements

The author and publishers wish to acknowledge the following photograph sources:

Aerofilms Ltd pp. 33 bottom, 61, 99, 103 bottom right, 108, 111 top, 112, 116
Cambridge University Collection p. 103 bottom left
Dr Michael Clarke, Department of Geography, University of Southampton p. 107 left
Forestry Commission p. 33 top
Keystone Press Agency Limited p. 63
Professor James P. Morgan p. 70
U.S. Army Engineers District, New Orleans p. 78
Most of the remaining photographs were taken by the author Keith Hilton.

The author and publishers wish to thank the following who have kindly given permission for the use of copyright material:

W.H. Freeman & Company for an extract from *Earthquakes: A Primer* by Bruce Bolt
Guardian Newspapers Ltd. for extracts from articles 'Town's Sinking Feeling' by Harold Jackson, and 'Canute takes on the Waves' by Michael Parkin.

The publishers have made every effort to trace the copyright holders, but if they have inadvertently overlooked any, they will be pleased to make the necessary arrangement at the first opportunity.

1 Landscape shaping

An Atlantic dawn

It was a grey November dawn in 1963, 35 km off the southern shore of Iceland. Olafur Vestmann, the cook of the fishing boat *Isleifur II*, was on watch. Below decks the rest of the crew drank coffee after finishing laying out their line. At 7.15 a.m. Olafur, feeling the boat make an odd twist, scanned the horizon and was amazed to see a cloud of dark smoke arising from the sea. Thinking it was a ship in distress the *Isleifur II* sailed closer. The crew then saw showers of ash, cinder and larger 'bombs' of solidified lumps of lava shooting out from a red glow in the sea. Beneath the dark cloud a sulphurous smell hung in the air. The crew were the first witnesses of the birth of a new volcanic island.

In Icelandic legend Surtur was a giant bringing destructive fire, so the new island was named Surtur's island or, in Icelandic, Surtsey. The birth of Surtsey and the efforts of water and wind to shape it during the first decade of its life is the main storyline of the chapter.

The chapter's theme: an introduction to geomorphology

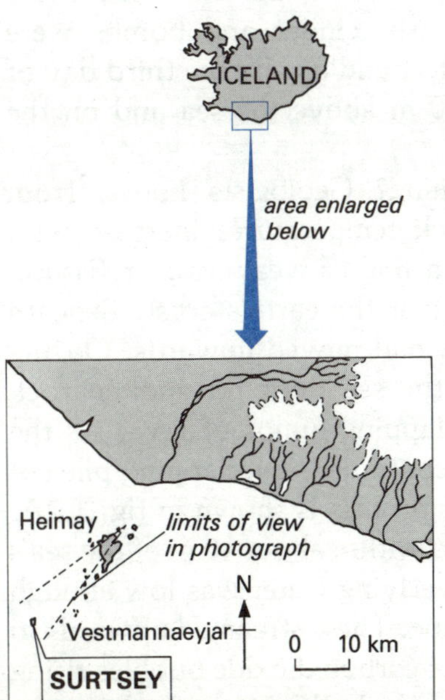

figure 1.1   *Surtsey's eruption viewed from Vestmannaeyjar in 1963. Ten years later this town itself was threatened by the eruption of Kirkufjell (just off the view to the left) whose ash and lava extensively damaged this fishing town and almost blocked its harbour.*

The landscapes you see at home, on your way to school or on holiday are complicated. After all they may have been produced by the effects of water (in all its forms), air and living things working over tens of millions of years. Today's landscapes might be compared to medieval manuscripts which were often partly scraped clean so that the precious parchment could be reused. The older words (i.e. the older landforms) can be partially read through the words which were written later (i.e. the landforms being produced by today's processes). In our present-day landscapes we can therefore find traces of features, now almost destroyed, which were produced millions of years ago. Another complication is the work of man. Landforms can be obscured by buildings, roads, mines and farms and the various landscape processes may be altered too.

For these reasons a new piece of land like Surtsey is a good place to begin our study of geomorphology. Later the scope of our study widens in space and time: to the Icelandic mainland and to aspects of the world's structure.

Surtsey: the story of an eruption

Birth

Surtsey grew where the sea was 120 m deep. The eruption, first seen at 7.15 a.m. by the crew of the *Isleifur II*, had by 10.00 a.m. produced a column of smoke and cloud 4000 m high. On the following day a small narrow ridge of black cinder appeared above the waves and the clouds of steam and dust reached more than 9000 m into the air! Meanwhile around the **vent** where the ash, cinder and bombs were escaping, a small **cone** began to build up. By the third day of the eruption this cone was 40 m above the sea and on the sixth day it was 60 m high.

What exactly was happening? Geologists know, from mining and drilling, that rock temperatures increase with depth. Where Surtsey grew, a line of weakness, or **fissure**, had formed in the upper part of the earth's crust. Beneath this, molten rock, or **magma**, had moved upwards. Oozing out through the fissures on the sea floor the molten rock cooled rapidly, forming overlapping lumps of **lava**. As the submarine volcanic eruption continued, overlapping piles of this **pillow lava** built up. This process is shown in fig. 1.2A.

When the pillow lavas got to within about 40 m of the sea's surface, the pressure of the overlying water was low enough for gases in the magma to escape. These streamed upwards to the sea's surface like the lines of carbon dioxide bubbles rising in a lemonade bottle when you undo the top and release the

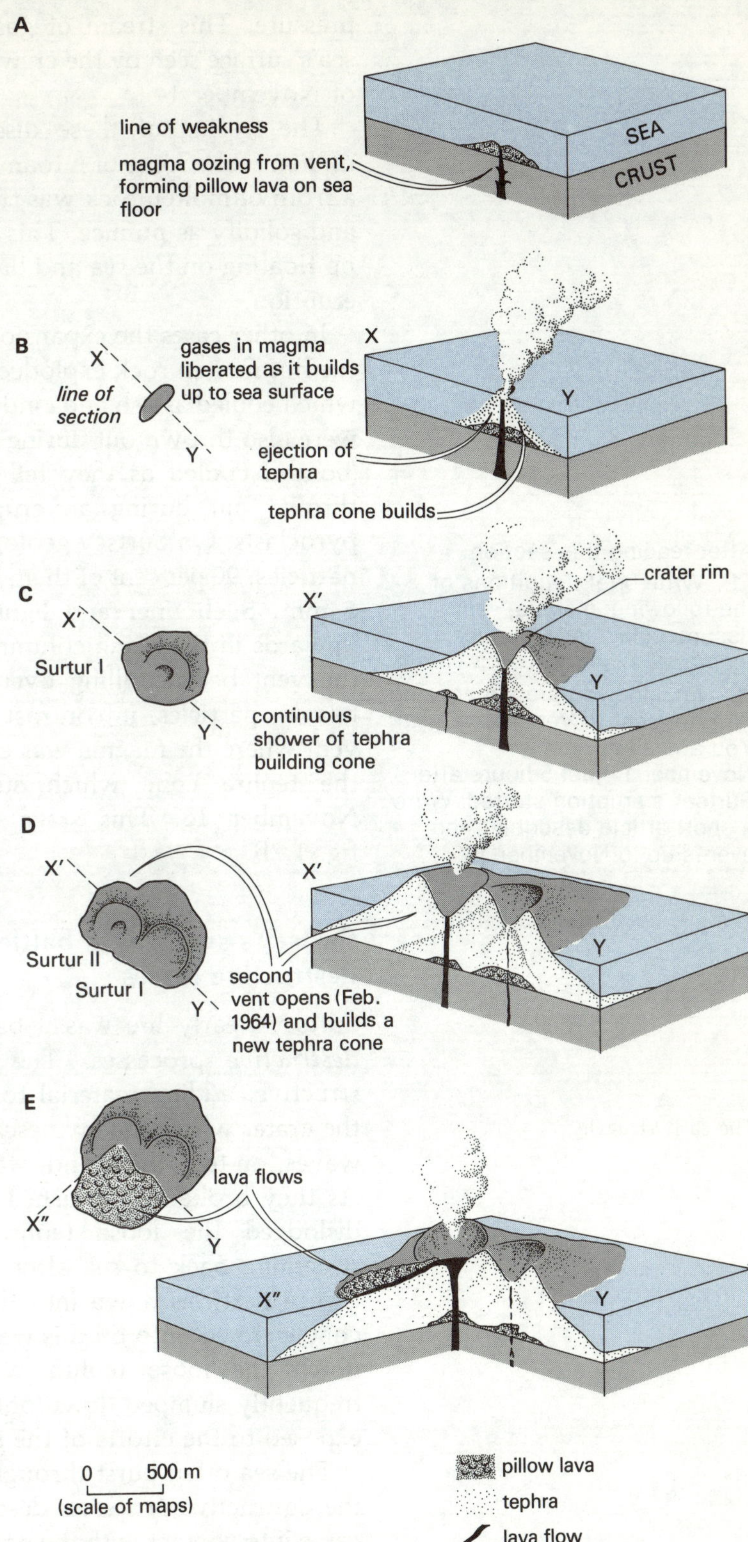

figure 1.2   The growth of Surtsey

pressure. This stream of gases caused the splashing of the sea's surface seen by the crew of the *Isleifur II* on the morning of November 14.

The freeing of these dissolved gases also affected the molten rock. Some of it foamed up as the gases expanded and a froth of molten rock was thrown out from the vent to cool and solidify as **pumice**. This was so light that it often ended up floating on the sea and drifted downwind throughout the eruption.

In other cases the expansion of the gases was so rapid that the hot molten rock exploded into a mass of small fragments which cooled as ash and cinders. Larger globs of molten rock were also thrown out during the eruption and these volcanic 'bombs' cooled as they fell through the air. The material thrown out during an eruption is known as **tephra** or pyroclasts. On Surtsey geologists measured the sizes of these particles: 90 per cent of them had diameters between 0.05 and 5 mm. Such finer and lighter dust particles were carried upwards in the rising column of steam and warm air above the vent before falling over a large area. The larger and heavier particles, in contrast, fell to the surface close to the vent where the magma was escaping. They piled up to form the tephra cone which surfaced above the waves on November 16. This stage in the eruption is shown in fig. 1.2B.

Surtsey's survival: a battle between constructive and destructive forces

Surtsey's early life was a battle between *constructive* and *destructive* processes. The volcanic eruption was constructive, adding material to build the tephra cone around the crater where the lava escaped. The Atlantic's winds and waves, on the other hand, were trying to destroy the island. As they broke and splashed up the shore the waves easily dislodged the loose tephra. The backwash of water, streaming back to sea after the wave had broken, clawed tephra particles down into the sea. The sides of the tephra cone were cut into by this wave attack. Above the wave-cut notch the loose tephra was now unsupported and it frequently slumped down onto the shore, where it was also exposed to the efforts of the sea.

The sea often burst through the crater walls as a result of the destructive processes described above. As the sea water came into contact with the hot magma in the vent it suddenly vapourised into steam. Whenever this happened a dull

After reading this section
 1   Write brief definitions of the following: magma, vent, fissure, pillow lava, pumice, tephra.
 2   Imagine you are a magazine reporter sent to Vestmannaeyjar. You arrived at noon on November 14 just 5 hours after Surtsey's eruption started. Write a short article describing the events up to November 16.

The early struggle

explosion was followed by a fresh uprush of tephra. The piling up of tephra continued during November and December 1963 but by January 1964 the crater walls had grown more substantial and secure. The explosions began to be replaced by a noisy dull roar as the uprush of tephra became more continuous.

A new eruption

In early February 1964, two and a half months after the *Isleifur II* saw the signs of Surtsey's birth, the original vent stopped erupting. A new crater opened to the north-west – Surtur II. From its vent a continuous tephra shower built new crater walls, reaching a height of 169 m. Figure 1.2D shows how the shape of the island became transformed by this new vent. The area of the island was now a square kilometre.

By April 1964, only five months after the eruption began, the sea was no longer breaking through into the crater – the supply of tephra had 'outweighed' the sea's ability to move it. The island had survived its first Atlantic winter. Scientists had begun to make landings to survey its growth and changes. As the island was still just a pile of loose tephra its longer term survival was by no means certain. The next stages in the eruption were going to be crucial.

The key to survival: The lava shield

Fortunately, during the following months a lake of red-hot molten magma gradually accumulated in the crater of Surtur II. Like wax driblets running down a candle, successive flows of glowing magma (at a temperature of 1110°C) overtopped the crater rim and flowed towards the sea. Geologists estimated that the lava was flowing at a rate 4 500 000 tonnes a day! The cooling and solidifying layers of lava veneered the loose tephra formed earlier in the eruption. Between July 1964 and May 1965 these lava flows built a shield, shown in fig. 1.2E, of about 1.5 km^2 over the southern half of Surtsey. Being more resistant to the sea's attack the growth of this lava shield marked another vital stage – Surtsey had survived its birth pangs and was now likely to be a substantial island.

Stillborn eruptions

Between May and October 1965 there was a new outburst of submarine volcanic activity 600 m to the east of Surtsey. Water spouting and tephra and pumice ejection heralded the birth of Syrtlingur, whose new tephra cone eventually reached 70 m above the sea. In December 1965 there was a second outbreak, this time 900 m to the south-west of Surtsey. This formed the island of Jólnir, which by August 1966 rose to a height of 70 m.

Syrtlingur and Jólnir formed just like Surtsey. (Remember the sequence of events described earlier and summarised in fig. 1.2.) Firstly lava oozed from a vent on the sea floor. Secondly pillow lavas built up. The magma was 'degassed' and the molten rock then bubbled and exploded upwards to pile up as mounds of loose tephra above the sea's surface.

These two islands were shortlived. Their loose tephra cones were never mantled with a harder lava shield as Surtsey was. They were therefore unable to resist the attack of the waves.

Meanwhile on Surtsey the volcanic story was not quite over. In August 1966 a 200 m long fissure opened across the eastern part of the island and a row of small craters formed along it, within the old Surtur I crater. Lava flowed eastwards from these craters to cover the south-eastern quarter of the island. This eruption petered out by the following summer and the last lava was seen flowing on June 5th 1967.

The eruption lasted three and a half years and had built an island of 2.8 km². As only a tenth of the material ejected actually extended above the sea the island's size did not reflect the true achievement of the eruption.

Surtsey's transformation

The shape of Surtsey after three and a half years of eruption is illustrated in fig. 1.3. The map shows sea depth so the stumps of Jólnir and Syrtlingur, mentioned in the last section, are seen.

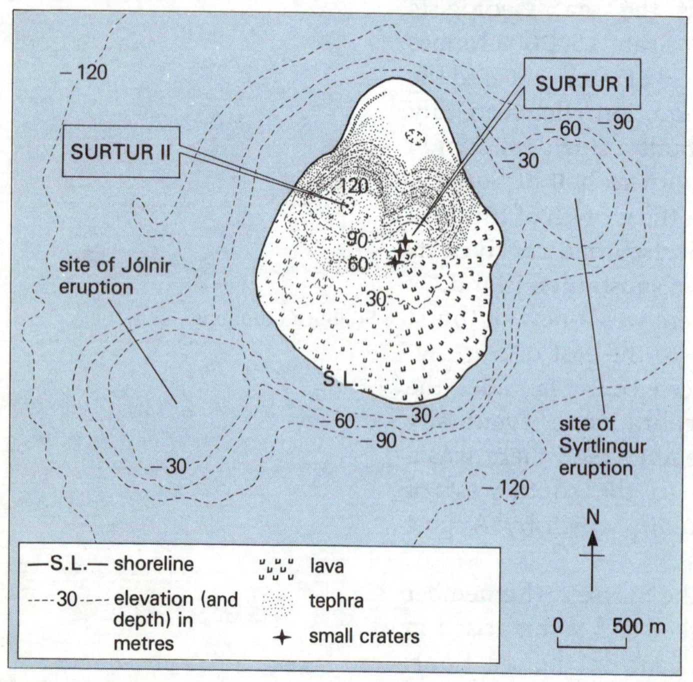

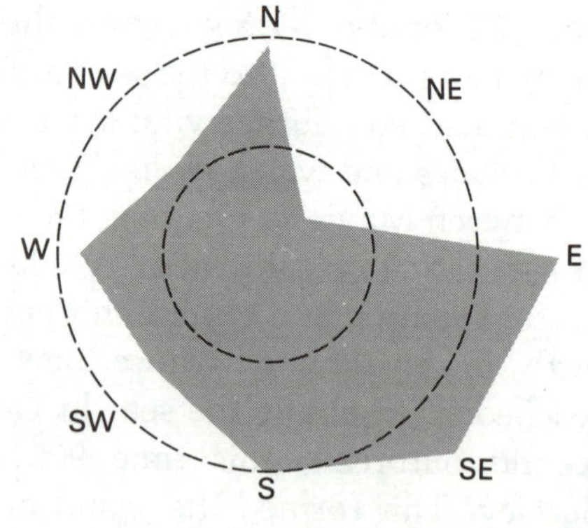

figure 1.3   *Surtsey in the summer of 1967*

During a volcanic eruption tephra thrown from the vent showers down around it. Most of this debris falls close to the vent, so a cone-shaped landform is built. Surtsey had two vents throwing out tephra and fig. 1.3 shows these cones overlapping.

Figure 1.3B suggests the reason for the asymmetrical cones. It is a diagram showing the frequency of winds from different directions (for Vestmannaeyjar, an island close to Surtsey, fig. 1.1). During the eruption tephra would have been blown away from the vent by the more frequent winds. It piled up more on the 'downwind' sides of the vent and produced the asymmetrical cones shown in fig. 1.3A. Therefore when the lava welled up (stage E in fig. 1.2) it rose to the level of the lowest rim (southern) overtopped it and spread to build the lava shield on the southern half of the island.

The volcanic processes were constructive, building up the island. Part of the *winds'* role in shaping the island has been described above but there were other agents working to alter it. The sea is one of these. The role of *waves* is especially important because they were at work from the island's first appearance. The outline of Surtsey in 1965 and again in 1967 is shown in fig. 1.4. The growth you can see in the south was produced by the lava flow, but there were changes in the shape of the island's northern tip. As you might remember, no lava spread in this direction and this triangle of low land was composed only of loose tephra (fig. 1.3A). If you look at fig. 1.5 the further changes in the six years after the eruption stopped can be seen.

The undercutting by waves has been mentioned earlier. This **marine erosion** is produced partly by the sheer pressure of water plunging and surging forward as the waves break. Sand, pebbles and even boulders are moved as the waves break on the shore. Surging backwards and forwards they act as tools to wear away or *abrade* the shore.

Just how much marine erosion takes place depends on the *size and energy of the waves*. This is influenced by three factors. First is windspeed: the faster the wind the higher and more powerful waves can become. Second is the length of time the wind has blown: it takes some time for large waves to grow.

The third factor we can discover from the diagrams. Figure 1.5 shows that the southern shoreline has retreated – at a rate of up to 30 m a year, in spite of the fact that this part of Surtsey was capped with the harder lava shield! The softer northern tip, on the other hand, has been reshaped with little overall loss of land.

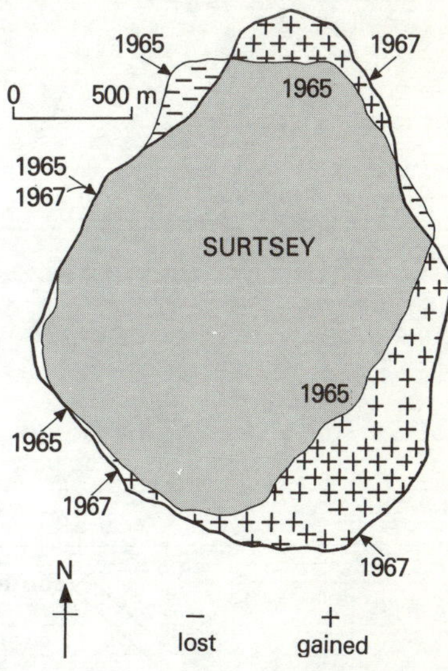

figure 1.4 Surtsey's changing shape 1965–67

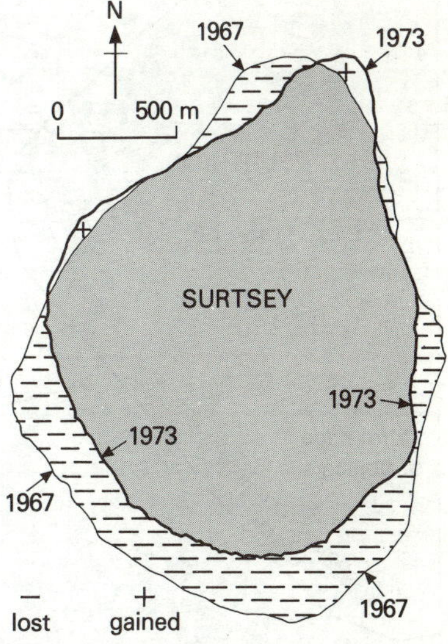

figure 1.5 Surtsey's changing shape 1967–1973

If you look at the map showing Surtsey's position (fig. 1.1) you might realise that the third factor affecting wave power is the 'exposure' of the coast. The distance which the wind is able to blow over the sea is known as the **fetch**. The longer the fetch the larger the waves can become. To the south of Surtsey lie thousands of kilometres of open ocean and the waves from the Atlantic can therefore be large and powerful. Not only is the fetch very long towards the south, west and east but the winds from these directions are also more frequent. The northern shore of Surtsey, in contrast, is exposed only to less frequent northerly winds blowing from nearby Iceland. These have a shorter fetch so that the energy of wave erosion is smaller from this direction.

The differences in wave energy around Surtsey are only a part of the story. To explain the changes in shape of the island's northern tip we need to look at the moving and building work of the waves.

With south-west winds the situation is as shown in fig. 1.6. On the north-western flank of the island waves approach obliquely. As they break, their swash travels up the beach in the same line as the waves' advance. The backwash, however, runs down the steepest slope of the beach (fig. 1.6C).

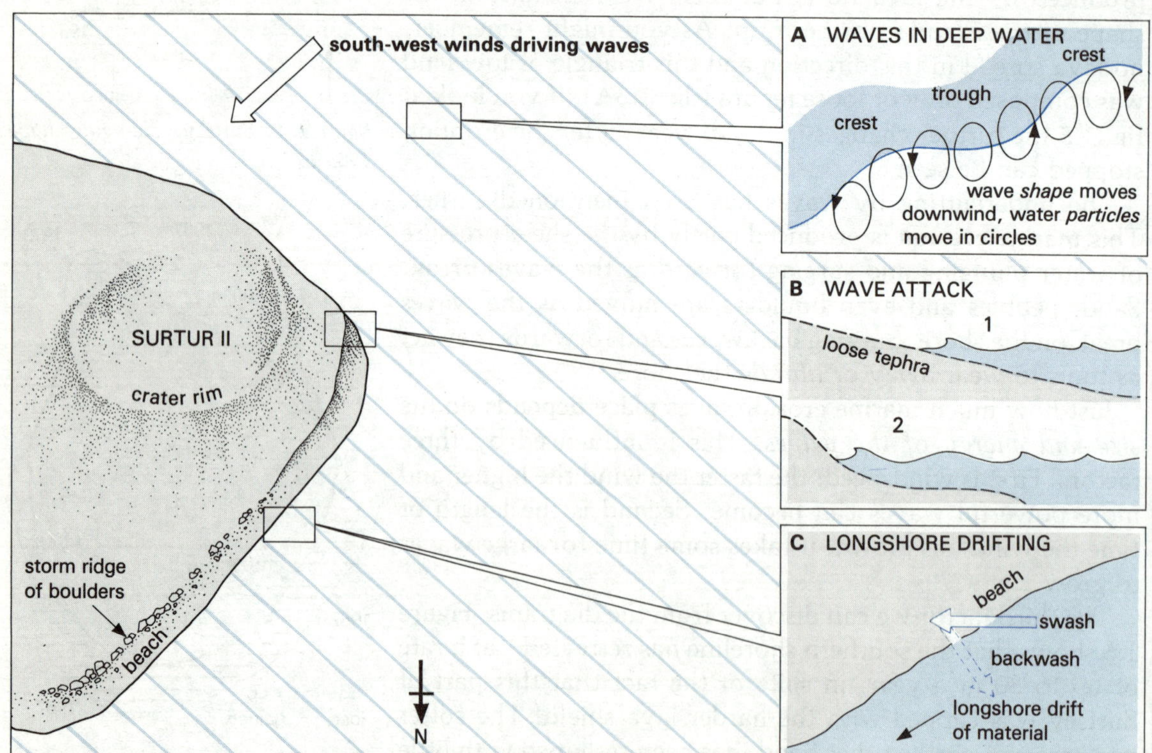

figure 1.6   Marine processes at work on Surtsey's western shores

Fragments of tephra picked up by the sea in the breaking zone of the waves, or on the beach, can therefore end up travelling along the shore in a series of zig-zag movements. This process is known as **longshore drifting**. With the south-westerly winds shown in fig. 1.3B, such drifting moves material towards the island's northern tip.

During storms, when wind speeds are higher, the waves are larger and more powerful. They have moved lava boulders up to 1.5 m in diameter to form the backbone of the northern headland's beaches. This 'storm ridge' is about 5 m above sea level. Behind it the loose tephra is sheltered and sand particles blown in by the wind are able to accumulate (fig. 1.3).

This moving and building also operates on Surtsey's eastern shore whenever easterly and south-easterly winds blow. Figure 1.5, however, shows us that the northern tip moved eastwards during a six-year period. This suggests that south-westerly winds and waves are more dominant.

What is happening is that some material eroded from the south and centre of Surtsey is being moved and used to reshape the northern headland. Such a triangular lowland built by two longshore drifting movements is known as a **ness**. The weak (low energy) and infrequent waves from the north are unable to blunt its tip.

Besides the work of the sea there have been other *agents of change* at work – the wind, rain and plants. Wind, which produced the asymmetrical tephra cones, continued its work after the eruption. The finer particles of ash have been winnowed out, or picked up, by the wind. They have then been dropped wherever and whenever the windspeed has fallen. They have thus accumulated in cracks and crevices of the lava surface and on the sides and in the lee of the craters.

The surface of Surtsey was at first just loose and porous ash (or tephra). At first the tephra was so porous that any rain falling onto it soaked (or infiltrated) into the debris. The molten rock cooled so quickly that individual minerals did not have time to form crystals, so that a 'glass' of basalt was formed. Exposed to air this slowly changed. The iron within it was oxidised and its other minerals, like calcium and sodium, were washed out by rain. In other words the basalt was **weathering**. It was becoming chemically changed. These were the first steps in producing less porous surfaces which could hold water.

Surtsey's colonisation by organisms fascinated biologists, Living things moved to the island in various ways. The sea washed buoyant and salt-tolerant plants from the shores of other islands and these were able to colonise the beaches. The

1 Examine figs. 1.4 and 1.5. Describe the changes in the shape of Surtsey 1965–1973.
2 Explain how wave energy varies around Surtsey's coast.
3 Why has Surtsey's southern shoreline retreated?
4 Describe how longshore drifting occurs and explain how Surtsey's northern tip has been formed.
5 On a copy of fig. 1.5 indicate what changes you would expect to find in the island's shape since 1973.

Weather and life

seeds of some plants were carried on the feet or in the droppings of migratory birds, like sedge carried in the gizzards of snow buntings arriving from Britain. Finally there was the wind which carried the light spores of mosses, lichens, bacteria and moulds.

In sand-filled hollows on the lava surface the bacteria, moulds, algae, moss and a few higher plants established *pioneer communities*. The first moss species was seen in 1967. By 1971 a moss carpet was beginning to develop on the lava shield and by 1973 some sixty-three species had arrived. This moss carpet collected wind-blown dust and the dead moss slowly accumulated as humus. This humus held moisture between the storms.

Life therefore started slowly on Surtsey. Its climate is similar to Iceland's – cool, windy and rainy (as warmer 'Atlantic' air alternates with colder 'Arctic' air, passed in the travelling depressions). For plants the initial problem was periodic dryness caused by the inability of the porous tephra to hold water between rainstorms. As the tephra weathered and as the first plants began to slowly improve the soil, so more demanding plants like sedge, crowberry, willow and red fescue grass could colonise. You may have seen similar processes of gradual plant *invasion* and the *succession* of more demanding plants on waste heaps or ground cleared for building near your home.

Widening the picture

This sounds familiar! In fact it is from St. Brendan's sixth century account of what is believed to have been an Icelandic eruption seen during his voyages. There is a long history of **volcanic activity** in Iceland and the Surtsey episode is only a small part of the saga. Since the birth of Surtsey an eruption has occurred on nearby Heimaey (fig. 1.1) in 1973 and at Krafta in 1981. Icelandic eruptions occur about as frequently as World Cup Soccer finals so by the time you read this there may well have been another.

Iceland is the most volcanically active area on earth. Figure 1.7 shows the recent major volcanoes. These volcanoes and eruptions include a range of different *shapes*, depending on firstly the type of lava erupted and secondly the kind of vent from which it escaped. Where the material ejected was ash (tephra) it has built volcanic cones like Hverfjall. When the lava is fluid it flows some distance before it finally cools and solidifies. If it escapes from a single vent such mobile lava can build a broad shield after a series of eruptions. Skjaldbreidur,

1   Make a list of the changes which occur when lava is exposed to the air.

2   Imagine you were the reporter in exercise 2, p. 4, sent back to Surtsey in 1975. Your editor has asked for a short, three-paragraph article covering the changes over the twelve years, to be illustrated by four photos. What would you write and what sort of photos would you try to obtain?

3   How do plants invade an area and change its soil?

'The whole island seemed one globe of fire and the sea on every side boiled up and foamed like a cauldron set on fire ... and a noisesome stench was perceptible at a great distance.'

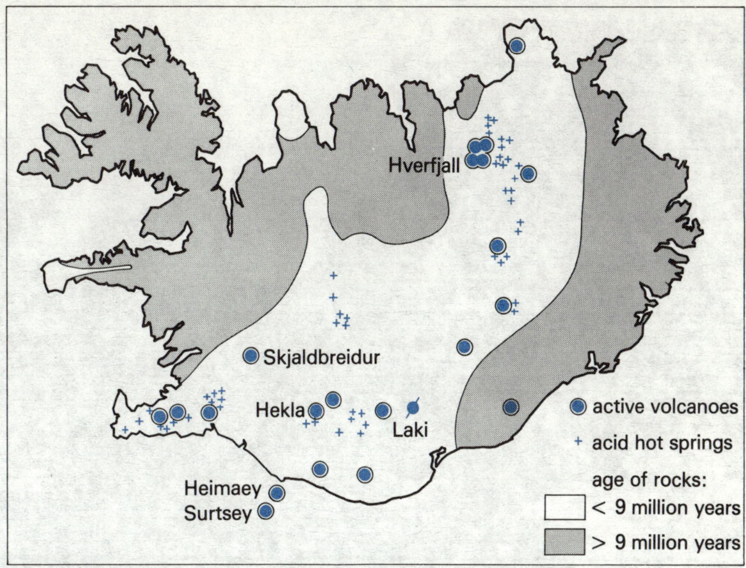

figure 1.7   Iceland's volcanoes

10 km across but only 600 m high and with gentle slopes of 7° or so, is an example of this kind of shield volcano. Composite volcanoes, like Hekla, are composed of alternate layers of ash and lava, often with several vents.

In some cases the lava escapes from a line of vents (a fissure). The Laki eruption of 1783–84 was of this type. The fissures ran for a distance of 25 km and the lava spread over 565 km² – an area larger than Greater Manchester! The ash from this eruption rained down on Iceland's pastures so that three-quarters of the island's sheep perished and in the resulting famine about twenty per cent of Iceland's people died.

In geological terms, however, volcanism is a creative process. It adds material to the earth's surface and tries to build new landforms. As we have learnt from Surtsey, these landforms are soon altered by the earth's surface processes.

A dramatic way to see the processes is to view an area from space. Figure 1.8 is a Landsat satellite image of part of southern Iceland, covering an area about 160 km along each side. Volcanism has been referred to as an important process in shaping Iceland's landscape. On the sketch map of fig. 1.8 you can see a fissure, a series of small cones along it, and a lava spread.

When you first looked at the image the areas which probably caught your attention were the lighter ones. The image was taken in August but these are areas of ice and snow. About 12 per cent of Iceland is permanently ice-covered and in the scene we can see parts of Vatnajökull,

11

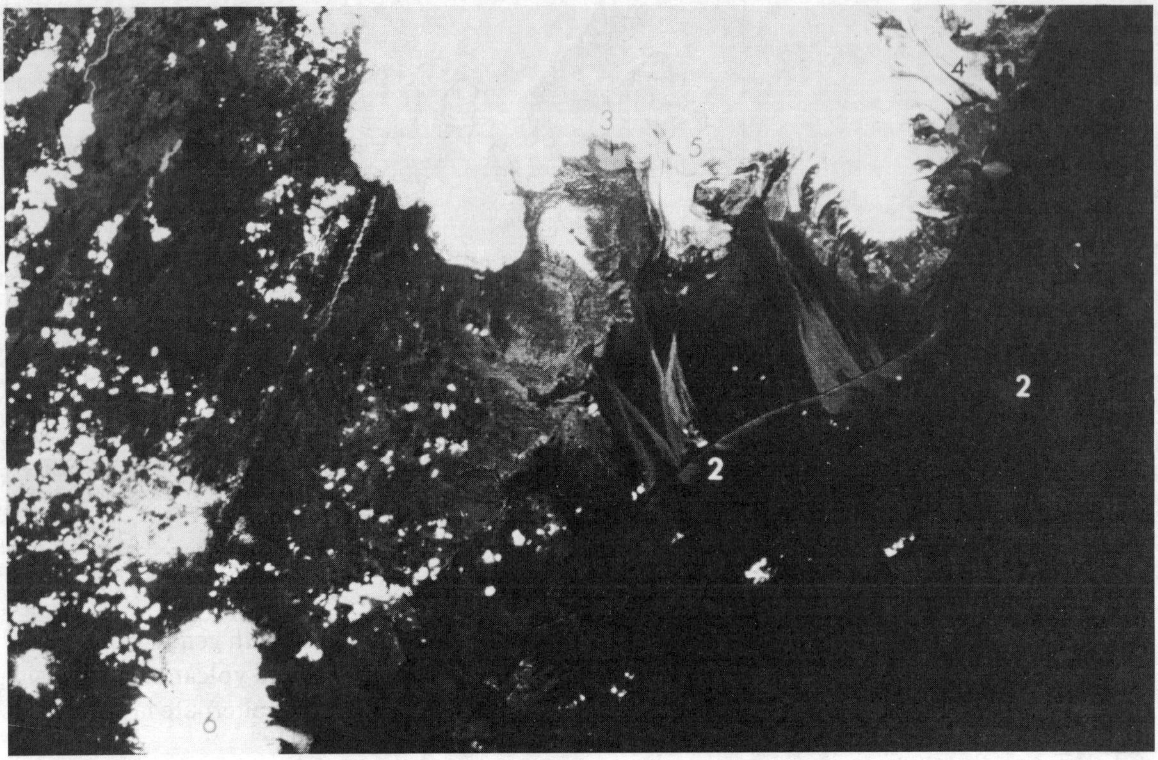

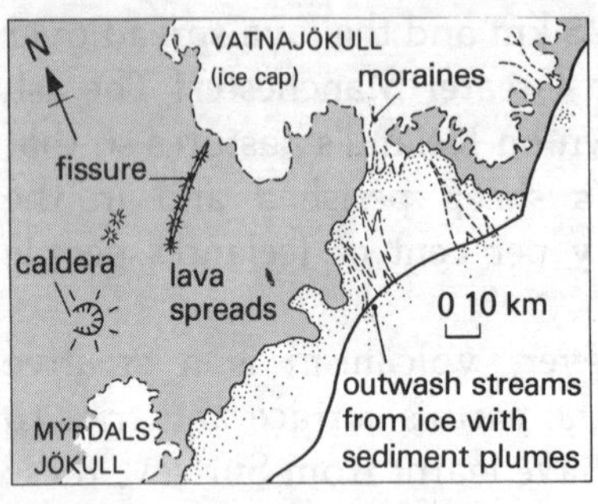

figure 1.8   A Landsat satellite image of southern Iceland

Erosion, transport and deposition processes are looked at in detail later in the book.

Iceland's largest ice cap, 60 km across and rising to a height of 2000 m. Part of the smaller Myrdalsjökull ice cap can also be seen.

Even from a height of 900 km signs of landscape processes can be seen. At the edges of the ice sheets streams are carrying the summer's meltwater away to the sea. The sea is discoloured by rock material being washed into it by these streams. These plumes of sediment are evidence of the **erosion** of the land by ice and its meltwaters.

A lot of the eroded material does not reach the sea. Instead it is deposited as great spreads of sand and gravel closer to the ice margin. These spreads of sand and gravel are known as **outwash deposits**. The sea itself of course is not idle. Along the coast, waves are eroding, moving and depositing material in the same kind of ways that they were on Surtsey.

There is one major agent of landscape change which was not stressed in the description of Surtsey – on Iceland the lava has hardened, weathered and become covered with a layer of plants and soil. A network of streams carry back to the sea the water which has fallen on Iceland as rain and snow. The streams also carry rock material. Some of this will have been eroded by the running water itself from the river beds and

12

sides, but a large part of it will have come from the slopes and valley sides. Some will be carried invisibly in the form of solution. The overall effect of running water on the landscape is the removal of rock material.

Frozen water (snow and ice) are also part of the cycling of water. Figure 1.8 showed that part of Iceland is covered by ice. Snow falling on higher ground does not melt away in summer. Gradually it thickens to form ice. This slowly moves downwards as glaciers to lower (i.e. warmer) elevations where it melts. Such moving ice is also a powerful agent of landscape change.

Iceland's place in the world's structure

If you imagine the earth as twenty-four hours old, Iceland would have formed in the last fifteen minutes, so it is young land. Figure 1.7 showed that on Iceland the most recent rocks occur in a belt across the centre. The centre of the whole Atlantic is also volcanically active. The eruptions of Jan Mayen in 1970 and Tristan de Cunha are examples. The sea depths in fig. 1.9 show the existence of a *mid-ocean ridge*. Only a few tiny parts of this underwater mountain range rise above sea level to form islands. Iceland is the only large land area to straddle it.

Volcanism is a sign of weakness in the outermost layer (**crust**) of the earth. Earthquakes too are another symptom of crustal movement. The distribution of both is shown in fig. 1.10. You can see that the earth, including the ocean floors as well as the land, is divided into two areas – those which are active (with belts of volcanoes and earthquakes as 'symptoms') and those which are quiet. The quiet areas are called **plates** by geologists. They are relatively stable areas of the crust; however, they move, so their edges are lively places.

Iceland straddles one of these plate boundaries: to the west the American plate, to the east the Eurasian. These are moving away from each other in the process known as *sea floor spreading*. Between them magma escapes, adding new material to the plates and building the mid-ocean ridge. This explains why the youngest rocks are found on the centre of Iceland (fig. 1.7) and why the further away you go from the mid-ocean ridge the older the rocks are. (Look at the ages in fig. 1.9.)

If fresh material is added at a plate margin, then somewhere else must be being destroyed or buried in the crust. This problem and some other aspects of plate movements are

On a tracing of the Landsat image (fig. 1.8) indicate some of the places where erosion, transport and deposition are taking place.

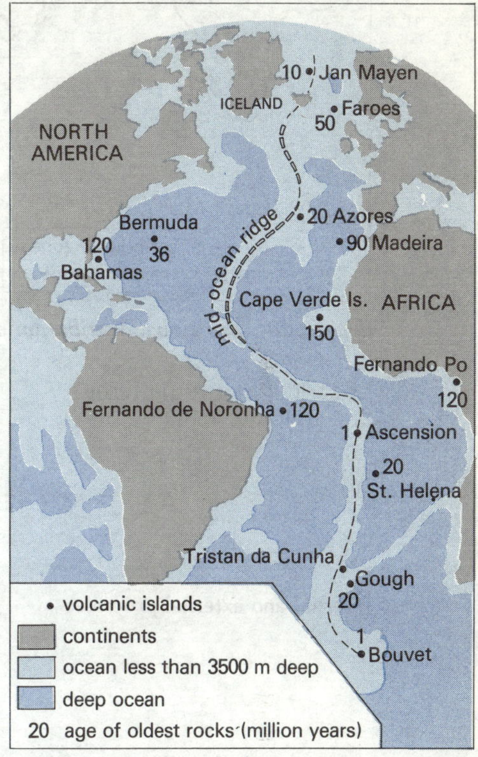

figure 1.9 The Atlantic: ocean depths and the ages of the volcanic islands

Crustal plates (discussed in chapter 8)

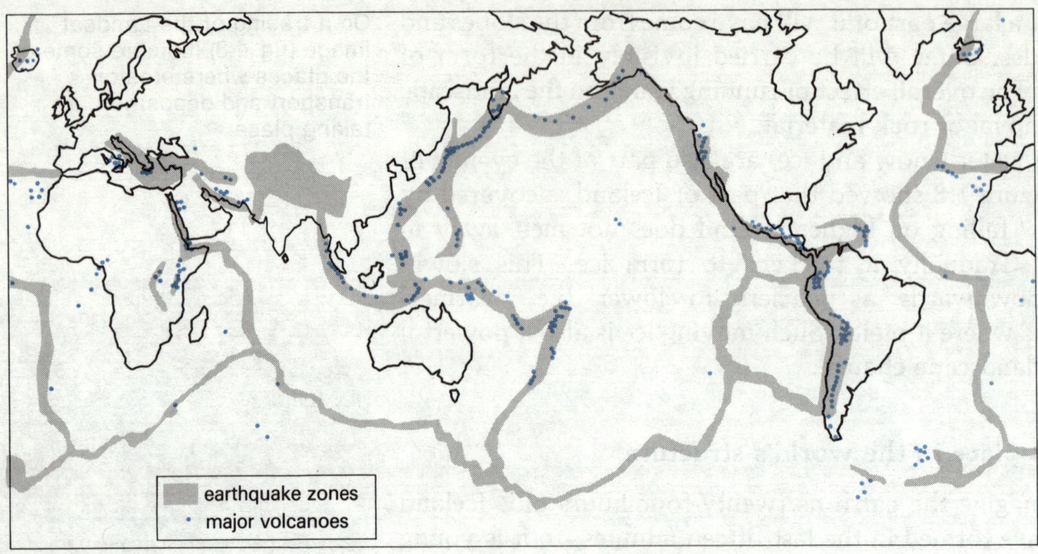

figure 1.10   The global distribution of volcanoes and earthquakes

Describe the two patterns
shown in fig. 1.10.

referred to in chapter 8. Before we leave this topic, however, it is useful to look at how rock materials (minerals) are cycled and moved.

How material and energy flows in the landscape

Internal and external processes

Landforms result from two processes – those internal to the earth (like volcanism) and those acting on its surface. The latter include weathering as well as the erosion, transport and deposition of rock materials by such agents as running water, ice and waves.

Energy

Volcanism, one input of energy from the interior, creates new landforms. Energy arrives from the sun too, cascading onto the earth to drive the hydrological cycle. Water, recycling to the sea across the land surface, helps to erode, transport and deposit rock material. The **hydrological cycle** is therefore of key importance in the shaping of the land. The sun's energy also 'fuels' the global wind systems which provide energy for waves and coastal processes. Finally the sun sustains life on earth and plants and animals also affect landscapes.

The hydrological cycle forms the
focus of later chapters

The summary diagram (fig. 1.11) also shows the rock *material* produced by volcanism being weathered, transported and deposited. All rock material on the land is on its way to the sea but it rarely makes this journey in one go. At any moment much of it will be 'in store', 'locked up' as debris on and beneath glaciers (as moraines), within river valleys, on slopes (as soils) or on the shoreline (as beaches).

The rock cycle

14

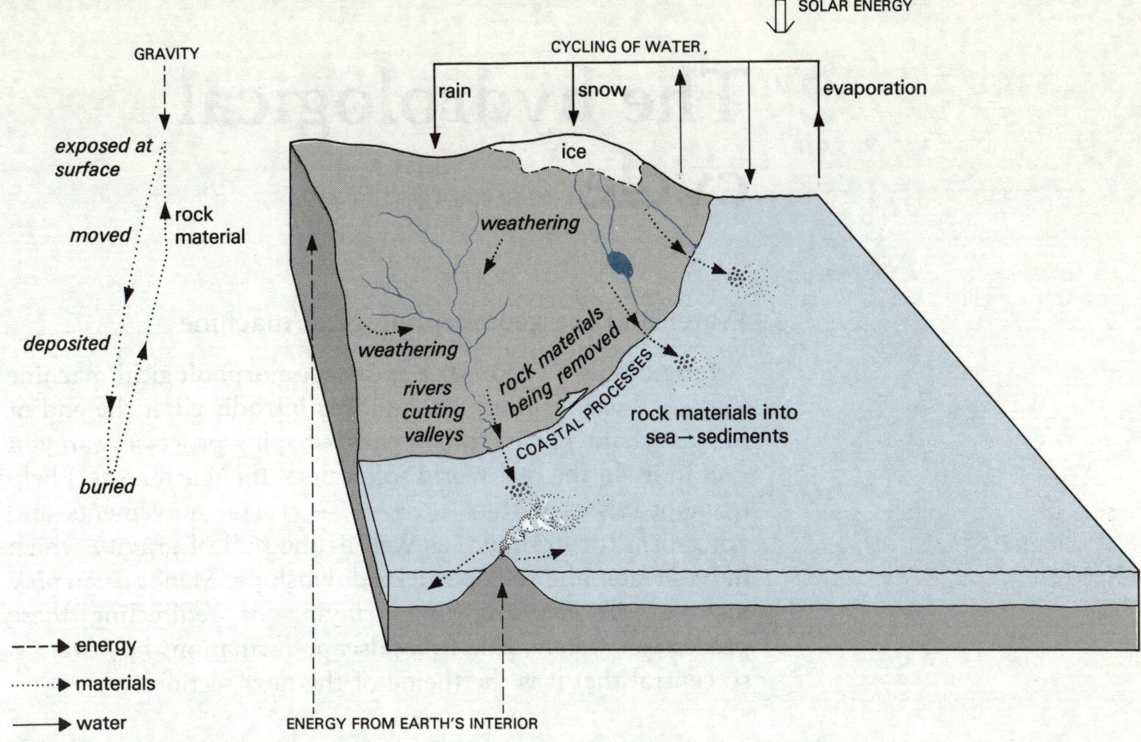

figure 1.11   Rock and water cycles

Glacial, fluvial (river) and coastal processes are described in later chapters

When it reaches the sea it can eventually accumulate to form sedimentary rocks. These may be later lifted above the sea (with energy from within the earth) to be attacked anew by the agents of landscape change.

Summary

Our story began on a tiny scale with the small island of Surtsey over two decades. It moved onto ideas about the earth over very much longer periods of time. What runs through the chapter are the ideas of landscape formation and change.

- Landforms are produced by internal and external forces.
- Processes act together.

2 The hydrological cycle

Water and the geomorphological machine

Arthur Bloom's vision of a great geomorphological machine (fig. 2.1) was inspired by the ideas introduced at the end of the last chapter. How many earth-shaping processes can you find in it? In the real world, of course, the machine gets help from processes within the earth – crustal movements and volcanism for example, as well as the pull of gravity which moves water and rock particles downslope. Man too can play his part in speeding up, delaying or redirecting these processes. Water's role in landscape formation, however, is so central that it is the theme of the next section.

figure 2.1   *The Geomorphological Machine (after Arthur Bloom)*

The world's water

To begin, let us look at the world's water. The oceans store the vast majority, about 97 per cent. The remaining 3 per cent, the world's *fresh* water, is held in the other **stores** shown in fig. 2.2. The largest of these is the water frozen in the Antarctic and Greenland ice sheets, the skim of pack ice on the Arctic Ocean and the glaciers found in the mountains of the world. The plants and animals of the biomass store hold only a minute amount of water.

Using the information in fig. 2.2 write a description of where and in what relative amounts the world's water is stored.

Transfers of water: the global cycling of water

Water continually moves between the various stores described above. These movements, or **transfers**, take place in several ways. Water moves between the stores in its three physical states, as a liquid, a gas or a solid. A **model** of the transfers in the global **hydrological cycle** is shown in fig. 2.3.

Like any other cycle there is no start or end; for convenience we will begin with the atmosphere. Water exists in the air as invisible vapour and as water droplets or ice crystals, which form clouds, rain and snow. Where has this water come from?

The proportion of the earth's surface area covered by oceans is 70 per cent. The **evaporation** of water from the sea's surface is therefore the most important source of moisture in the atmosphere. On land, water can evaporate from lakes, swamps and rivers as well as from the surface of the soil. Plants, too, draw up water from within the the soil and pass it into the atmosphere by the process of **transpiration**. Transpiration can move surprisingly large amounts of water

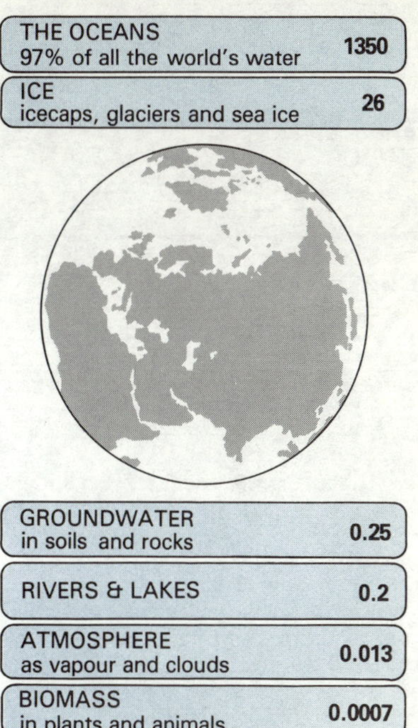

figure 2.2   *The world's water. The stores of salt, fresh, frozen and atmospheric water.*

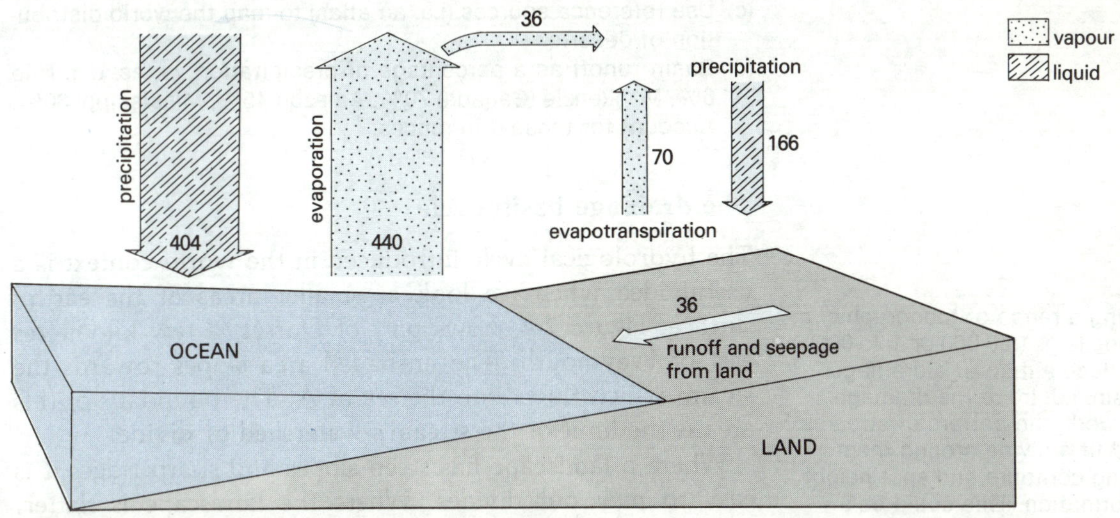

figure 2.3   *The transfers of water in the global hydrological cycle*

– a deciduous woodland in summer can transpire as much water in a day as evaporates from a lake of the same area. **Evapotranspiration** is the term used for both processes which move water from sea, soil and plant 'stores' to the atmosphere 'store'.

Once the water is in the atmosphere it can move – upwards or sideways. Before, during or after moving, vapour may be condensed into visible droplets, or if the air is cold enough into ice crystals. Under certain conditions the minute cloud droplets and the ice crystals become so large and heavy that they fall to the earth's surface as **precipitation**. This is the term used for all the different forms of this transfer process i.e. rain, drizzle, sleet, snow and hail.

If the precipitation falls into the sea the evaporation-condensation–precipitation cycle of the water molecules is simple. However, if it falls on the land alternative routeways exist. The water can either return to the atmosphere from the soil and vegetation by evapotranspiration (mentioned above) or travel by way of streams and rivers to the oceans. This second transfer, the **runoff** of water from the land, is a key process in landscape shaping.

1 List the terms used for the transfers of water in the hydrological cycle.

2 In fig. 2.3 the arrow thicknesses reflect an 'average' year and an average earth, hardly very realistic! Look at the right-hand end of the figure.
(a) In 'humid' environments the climate produces a surplus of precipitation over evapotranspiration losses. Draw a version of fig. 2.3 to show this.
(b) In arid environments (deserts) precipitation is lower and evapotranspiration losses higher. Sketch a version of this situation.
(c) Use reference sources (i.e. an atlas) to map the world distribution of deserts.
(d) Basin runoff as a percentage of precipitation varies, e.g. Nile 6%, Mackenzie (Canada) 70%, Amazon 45%, Mississippi 30%. Account for these differences.

The drainage basin cycle

The hydrological cycle introduced in the world context is a useful idea when we look at smaller areas of the earth's surface. Figure 2.4 shows part of Dorset, a few kilometres east of Weymouth. The unshaded area slopes towards the steam, which flows into the sea at A. The boundary of this area is the limit of the stream's **watershed** or divide.

Where a landscape has steep slopes and sharp ridges it is easy to pick out divides. Where the landscape is flatter, identifying a stream's divide may not be so easy. Checking in

Using a range of topographic maps (e.g. 1:50 000 or 1:25 000) produce a map of a drainage basin, i.e. trace the drainage network, the pattern of streams, and fit a divide around them using contours and spot height information. This could be a basin close to your school.

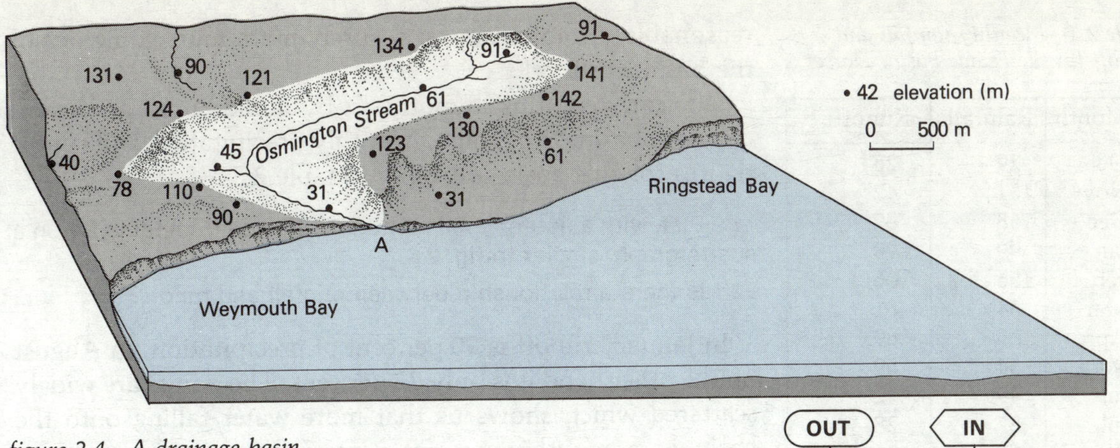

figure 2.4   A drainage basin

the field may then be needed to discover very gentle gradients down which water may flow. The land enclosed by a stream's divide is known as its **drainage basin**. The basin shown in Fig. 2.4 is quite small, with an area of less than 4 km². At the other extreme the Amazon drains an area of 5 775 000 km²! The earth's land surface is covered in a mosaic of drainage basins of different sizes and shapes. As working units of the landscape geomorphologists find them a central area of study.

Building a model of the drainage basin

Figure 2.4 is a block diagram of a drainage basin. In fig. 2.5 the basin has been simplified as a black box, with precipitation entering it. This either ends up as runoff, or returns to the atmosphere by evapotranspiration. These form the outputs from the box. This input–output diagram is a **model**. During the rest of the chapter we will try to 'open up' this black box.

Models in geography take many forms: diagrams, maps, graphs, equations, computer programs or even hardware. By simplifying the real world, models can help us understand how it works. Sometimes they help us with 'what if' questions. We could use our simple black box model as an example. If there is more input we could expect more output. This idea could be stated as a **hypothesis** (a statement of expected relationships): increased runoff will result from increased precipitation. If this hypothesis is correct a scattergraph of precipitation (input) plotted against runoff (output) would look like fig. 2.6. The graph shows that as precipitation increases so does runoff. (This is a positive relationship between two variables.) This hypothesis sounds

figure 2.5   An input–output model of the Drainage Basin (Stage I)

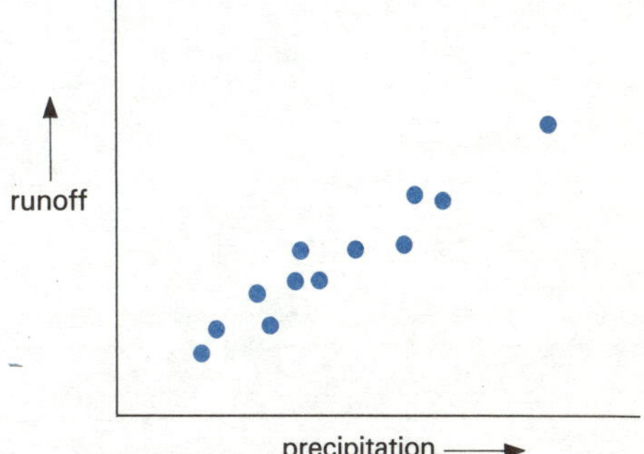

figure 2.6   Rainfall and runoff relationship which would prove hypothesis 1

19

Month	Rainfall	Runoff
Oct	29	25
Nov	151	42
Dec	168	87
Jan	99	90
Feb	156	106
Mar	34	81
Apr	138	69
May	64	59
Jun	49	42
Jul	57	28
Aug	118	26
Sep	27	21

1965–66 water year. Figures in mm

To explain how water moves in the drainage basin is the aim of this section. It introduces terms for the various water movements and stores. Understanding these terms is important for the next chapter.

reasonable: if more rain falls, surely more must come out of the basin as runoff.

We can test this hypothesis by looking at an actual drainage basin, the Frome. Monthly precipitation and runoff figures for this basin are given in table 2.1.

1 Plot, with a dot for each month, the figures from table 2.1 on a scattergraph, similar to fig. 2.6.

2 Is there a relationship between rainfall and runoff?

In January runoff is 90 per cent of precipitation, in August on the other hand it is only 22 per cent. The points are widely scattered which shows us that more water falling onto the basin in a month does not necessarily mean more water leaving it as runoff. In this case, our hypothesis is not proven; there is no simple relationship between precipitation and runoff.

Our simple model of a drainage basin needs refining. We need to consider the other output from the model (evapotranspiration) and we also need to look at how water moves within the basin.

The pathways of water: building a systems diagram of the drainage basin hydrological cycle

If you flew over the land you would see pastureland, cropland, forests, roads, buildings etc. The surfaces of drainage basins are therefore varied. Part of the surface is the network of streams and rivers. Some of the precipitation falls directly onto them. Figure 2.7 has therefore 'stripped back' some of the black box to show this. The precipitation is being added to the 'channel store', the streams, before leaving the basin as runoff.

Maps, however, tell us that streams occupy only a tiny proportion of the basin's area. We need to ask ourselves what happens on the rest of the basin. In humid environments* the land surface is blanketed with plants which can have quite an influence on the incoming precipitation. Imagine the start of a rainshower. Some water may get trapped on the leaves and branches of the plants. In a short rainshower almost all of the precipitation may get held up in this way. When the shower ends it will evaporate back to the atmosphere, having never reached the ground. If the period of rainfall is longer, once all the surfaces of the plants are covered with droplets of water,

* Humid environments are those where the precipitation exceeds evapotranspiration, in other words where there is sufficient water to keep streams flowing during the year.

any extra starts dripping through the leaves and running down the stems. The process of 'holding back' some of the incoming precipitation is known as **interception**. It is illustrated diagramatically in fig. 2.8.

How much precipitation is 'held up' depends on a number of factors. One, storm size, was mentioned in the last paragraph. Another is the *type of vegetation*, its density of foliage and stems. Caught in a rainshower you would be foolish to shelter under a birch tree if a sycamore was close at hand. The small leaves and fine branches of the birch are not nearly so effective in intercepting the rain as the large leaves and branches of the sycamore. Interception can also vary *seasonally* as the vegetation goes through its yearly growth cycle – neither the birch nor the sycamore would offer much protection in winter.

We have begun to build a picture of how water arrives at the land surface in the basin (fig. 2.8, right-hand end). How much takes each routeway depends on a range of factors: the type of precipitation, the season and the nature of the plant cover in the basin.

The next question is what happens to water when it reaches the surface. The transfer process of water seeping into the ground is known as **infiltration**. Figure 2.9 A shows how the

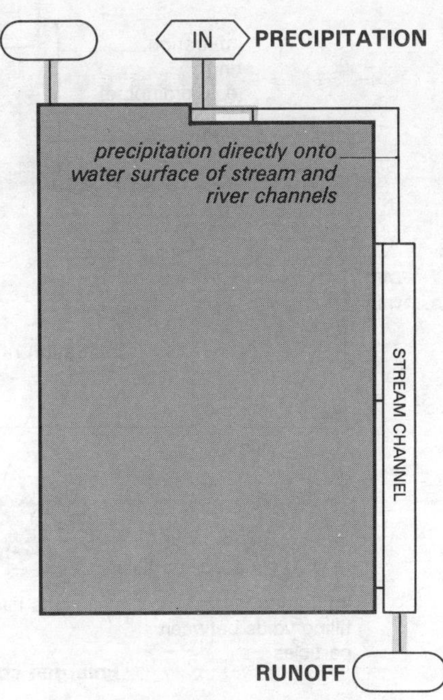

figure 2.7   The Drainage Basin Model, Stage II: incorporating channel stores (streams)

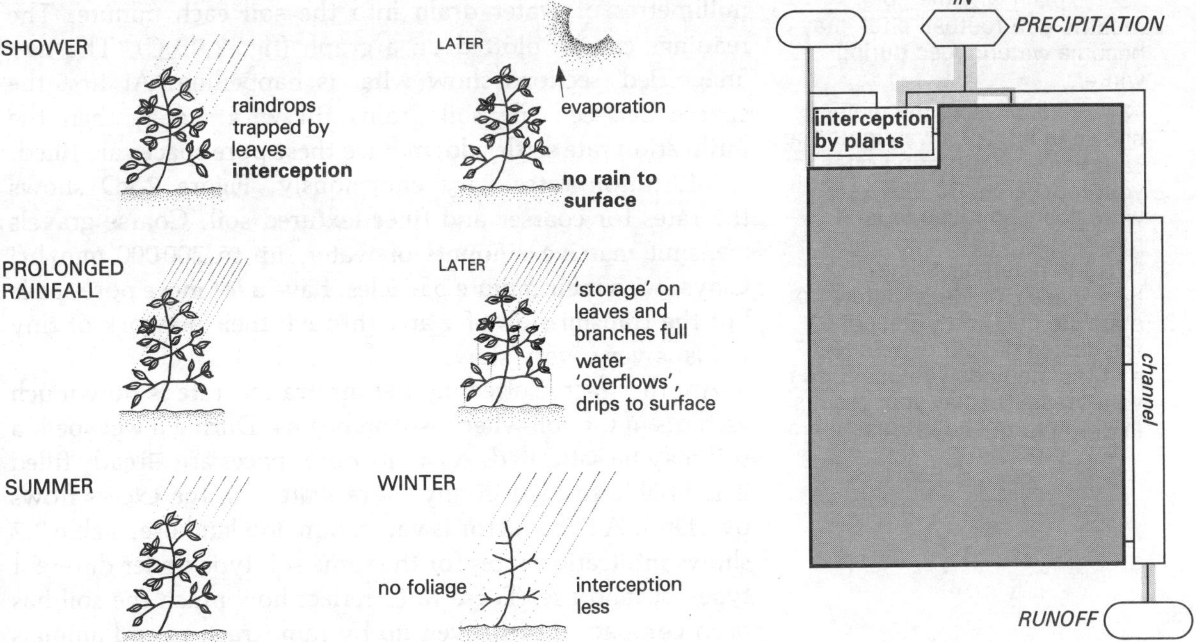

figure 2.8   The interception of incoming precipitation by plants and the Drainage Basin Model Stage III. The model shows water falling directly on streams, the soil and vegetation (where water is temporarily stored before going back to the air or transferred as stemflow and drip to the surface).

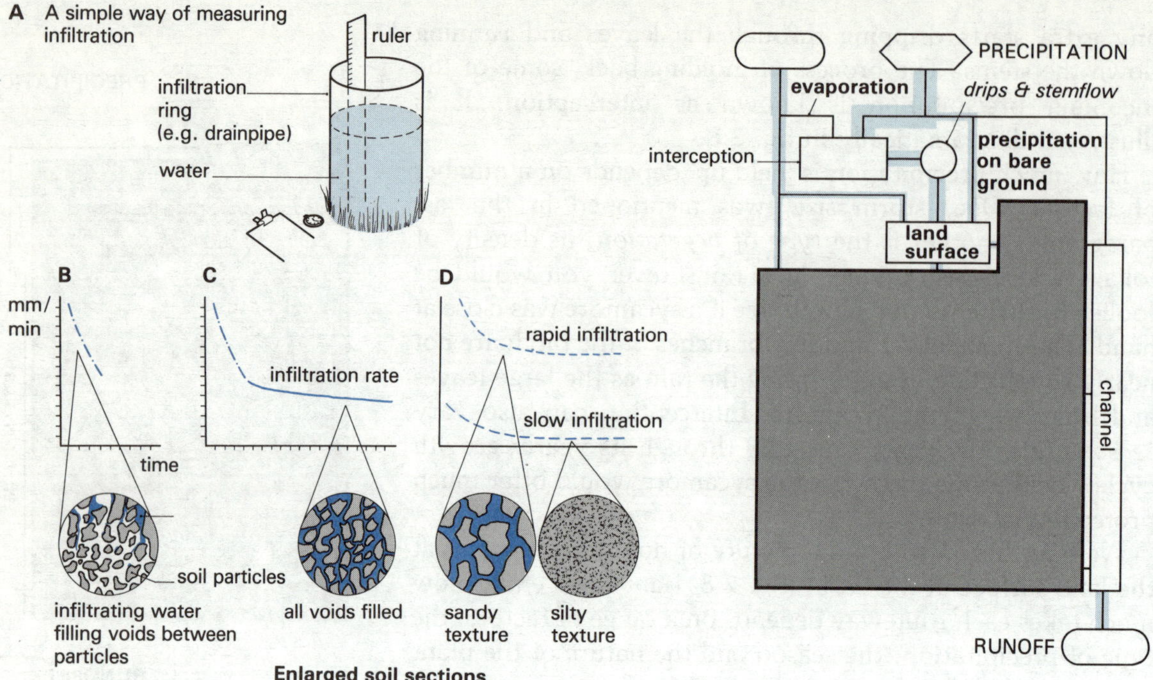

figure 2.9   *Soil infiltration: measurement and variables*

1   Describe how infiltration could be measured.

2   Explain why the goal mouths of a football pitch may become waterlogged during winter.

3   *Project.* Using the method shown in fig. 2.9, investigate the variations in infiltration rates in your local area. Do they vary with slope, soil type or land use?

Try to establish some hypotheses, or ideas, before you start the field work. Describe the design of your experiment and the methods you use to test your ideas. Display your results in graph form and discuss them.

infiltration rate of a soil can be measured by driving a pipe into the ground, filling it with water and noting how many millimetres of water drain into the soil each minute. The readings can be plotted on a graph (fig. 2.9B,C). The two 'magnified' sections show what is happening. At first the spaces between the soil grains fill with water, then the infiltration rate settles down once these pore spaces are filled.

Infiltration rates vary enormously. Figure 2.9D shows the rates for coarser and finer textured soil. Coarse gravels transmit massive amounts of water, up to 200 000 mm/hr! Clays, with their minute particles, have a lot more pore space but the transmission of water through their network of tiny voids is very, very slow.

Another factor affecting the infiltration rate is how much water is in the soil when a storm begins. During a wet spell, a soil may be saturated. As all its pore spaces are already filled it is unable to take in any more water, so the excess flows overland. A third factor is vegetation and land use. Table 2.2 shows infiltration rates for the same soil type under different types of land use. These rates reflect how much the soil has been compacted or broken up by rain, tractors and animals and how well the roots have penetrated the soil.

If rain arrives faster than the rate it infiltrates the ground, water begins to accumulate on the surface, on flat ground as

22

puddles. On slopes this surplus forms a surging film of water flowing *overland* towards lower ground. At the end of the storm this surface water may evaporate back to the atmosphere and take no further part in the basin's processes.

Once the water is in the soil three alternatives exist: it can move up, down or sideways. It may go back to the atmosphere (i.e. 'upwards') by evaporation and transpiration as plants draw water from their root zones. How much moves this way depends on the climate and vegetation. Some plants demand more water, some are just 'thirstier' in their growing seasons. Some climates, like those in the arid zones (especially deserts), have pronounced moisture deficiencies.

The water in the soil is trying to be pulled by gravity further down into the soil, eventually seeping into the bedrock. Such vertical seepage is often not very quick; the soil and rock textures may slow the downflow so that water may start moving in our third direction. Such *sideways* or downslope movement through the soil itself is known as **throughflow**.

The water which seeps downwards into the rock enters the 'groundwater store'. How fast it moves and how much water can be stored in the pores, cracks and fissures of the rock depends on how many and how big these spaces are. The amount of water stored is huge when it is compared to the other stores in a basin, such as the interception (plants), surface (puddles) and soil zone stores. Water from the groundwater store seeping into streams is known as **baseflow**.*

Summary

- The *pathways* of water in the basin have been described and defined in the last section. The drainage basin has been simplified into an input–output *model* (fig. 2.10). This diagram represents the basin as a *system*, a series of parts connected together.
- How drainage basins work in the real world depends on how fast the water moves and how big the stores are. How fast water can move is shown in fig. 2.10's 'speedometers'. Actual speeds are important to us when we try to control flooding. You might contrast the *quick* processes of channel flow, overland flow and throughflow, on the one hand, with the *slow* groundwater flow on the other. How these affect river behaviour is looked at in chapter 3.

* baseflow is discussed in the next chapter

table 2.2   Variations in infiltration rates with different land uses on the same soil type

Land use	Infiltration rate (mm/hr)
Old permanent pasture	57
Heavily grazed permanent pasture	13
Grain	9
Bare ground	6

1   Draw bar graphs to show the variations in infiltration in table 2.2

2   Suggest reasons why:
(a) cattle grazing can reduce infiltration;
(b) bare ground has such a low rate compared to the old permanent pasture.

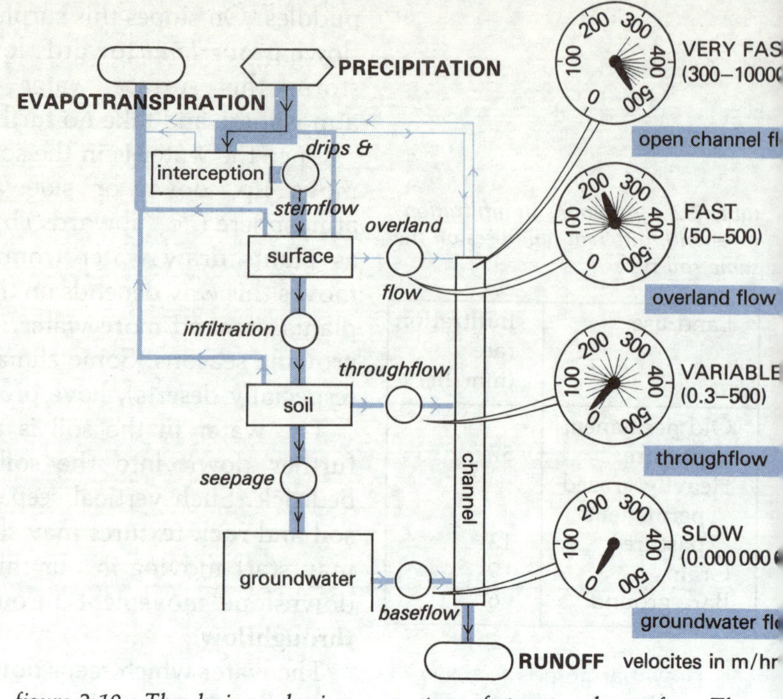

figure 2.10 *The drainage basin as a system of stores and transfers. The velocity of transfers is also indicated by the speedometers*

- Not mentioned in the chapter is the role of man in altering the routing of water. We can affect interception by altering vegetation – clearing forests and replacing them with croplands for example. We can affect the soil –altering the infiltration 'valve', by ploughing it, by letting farm animals walk on it or by covering it up with roads and buildings. Thirdly, we can affect the *channel* itself – by deepening, straightening, widening or damming it. Man as a farmer, at all levels from subsistence to intensive commercial farming, has therefore influenced the basin cycle. Similarly, as a civil engineer he has increasingly tinkered with the workings of the basin system as he 'messes about' with rivers.

Exercises

1 On a copy of fig. 2.4 (the block diagram) describe the pathways of water in the basin. Mark in the inputs, stores, transfers and outputs of water.

2 Using fig. 2.10B to organise your ideas, describe the quickest and slowest pathways of storm rainwater through the various stores and transfers before leaving the basin as runoff.

3 The drainage basin hydrological cycle

Why bother to understand the system?

Two views from the same spot on the banks of the Caroni River in Venezuela are shown here. One was taken in the wet season, the other in the dry when you can see that the river has shrunk dramatically and exposed much of its bed. These are an illustration of seasonal runoff variations, but in this chapter we will also be looking at changes over a few hours as well as those covering decades.

figure 3.1   *The Caroni River, Venezuela, in the dry season*

figure 3.2   *The Caroni River, Venezuela, in the wet season*

Understanding rivers is important. After all, we use them as sources of water for ourselves, our animals and crops. We often use rivers for transport, for power generation and for recreation. In societies as different as the Chinese Han dynasty of the third century BC and twentieth century America (in the case of the Tennessee Valley), people have tried to control or manipulate the flows and stores of water in the basin system – in other words to manage the basin hydrological cycle in order to exploit water as a **resource**.

Floods at one extreme, and low flows in droughts at the other, are also part of the picture. These are **hazards** for society. Of all natural hazards (earthquakes and tsunami, volcanoes, hurricanes, avalanches, etc.) river flooding is the most damaging to life and crops. If we wish to reduce the impact of hazards and improve the management of water as a resource it is vital that we understand how the basin system works.

Runoff variations

Figure 3.3 shows the daily output of water from a basin under three contrasting conditions. This small Dorset basin was the one used when the idea of the drainage basin was introduced. An 'average' day (if you can imagine that under British weather conditions!) would see about 9000 m³ of rain falling over the basin area. The 'average' runoff amounts to 4400 m³ a day. In May 1979 a hundred times as much water left the basin after a very heavy rainstorm. Figure 3.3C, in contrast, shows the runoff in a dry spell. Even in England daily runoff can vary enormously.

A storm in a small basin

Figure 3.4A is a map of the area around the Slapton Field Studies Council centre in Devon. It shows the basin in which students visiting the centre have taken measurements.

The basin's input and output, rainfall and stream behaviour, are basic observations. Conventional rain gauges collect the rain which has fallen between visits to the gauge, i.e. six or twenty-four hours. For short-term variations these

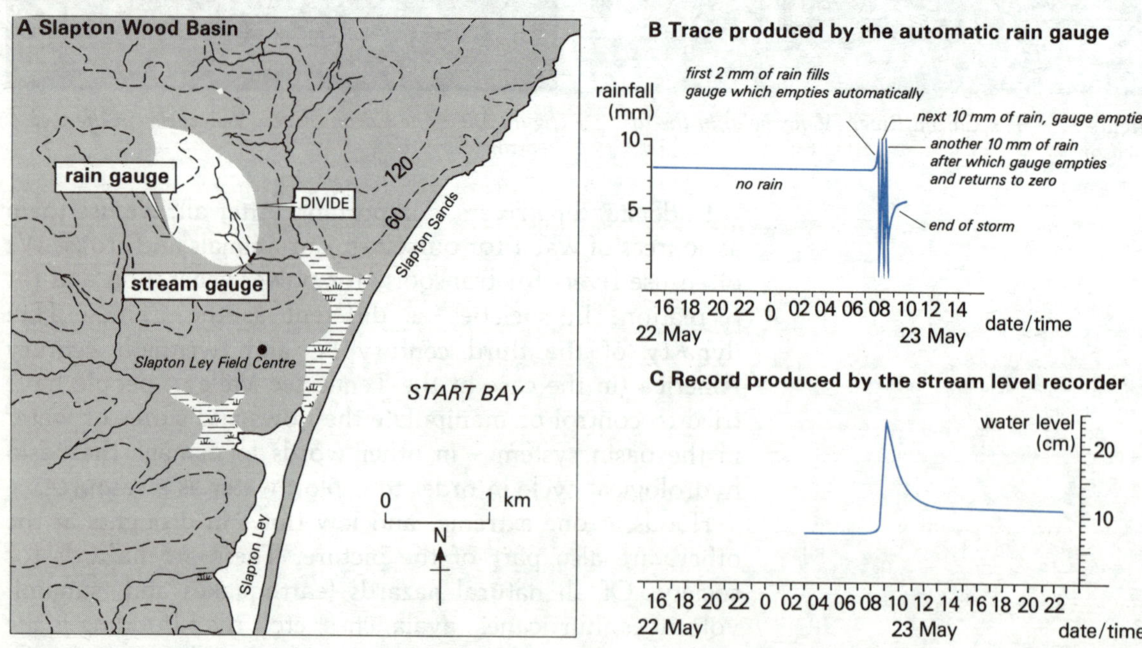

figure 3.4   The Slapton Wood catchment, Devon: rainfall and stream behaviour during a storm in May 1975

Beside the left margin:

figure 3.3   Three daily outputs from the Osmington drainage basin. The symbols are proportional to the volumes of water leaving this basin, which was shown in fig. 2.4.

A Average   B Flood   C Dry spell

Measurements of Inputs and Outputs

1 Look at fig. 3.4B. Describe how much rain arrived between 0700 and 0900 on May 23rd.

2 Look at fig. 3.4C. Describe when the stream level began rising, at what time it peaked and how it declined. Compare it to the storm's rainfall shown in fig. 3.4B.

A The depth zones on the slope

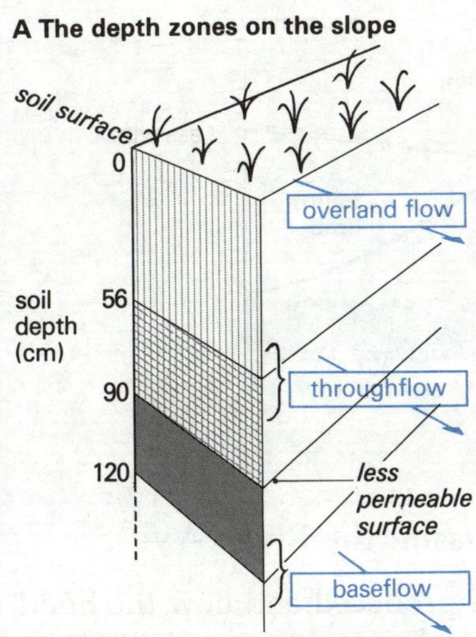

B Water arriving from various zones in the soil

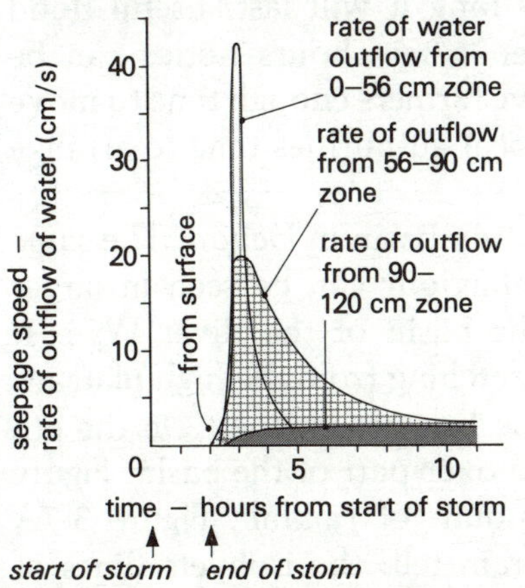

figure 3.5   Water seeping from soil on a slope

may be too crude and an automatic gauge may be more useful. Figure 3.4B is an example of the record from such a gauge at Slapton. It tells us that a total of 70 mm of rain fell in a period of only seventy minutes during an August storm. Automatic gauges like this can therefore tell us when and how much rain fell so that the **intensity** of rainfall (mm/hr) can be discovered.

The other basic observation is of basin output. One way to measure output is in terms of stream **discharge** (often expressed as so many cubic metres of water per second, or **cumecs**). To do this, the stream's speed can be measured, the cross-sectional area determined and the discharge calculated. Alternatively, a continuous record of the depth of the stream can be made. Figure 3.4C is a record of water level of the Slapton stream before, during and after the rainstorm shown in fig. 3.4B. Its sharp rise and slow fall is typical of stream behaviour in response to a storm. Notice that the highest water level occurred after the start of the storm. In other words there is a time *lag* before the stream (output) responds to the storm (input). There is also a gently falling 'limb' to the trace of water level once the rain has stopped. Our study of water moving through the basin in the last chapter (fig. 2.10) helps us explain why the stream behaves in this way.

Water which falls into the stream itself can move rapidly (fig. 2.10) and is the first storm water to arrive at the stream measuring station. As streams cover a tiny part of the basin area (often 1 per cent) the amount of water involved is quite small, but it contributes to the first steep rise in stream level. It is what happens to the water falling on the remainder of the basin which will help us explain the rest of the stream's behaviour.

The first few millimetres of rain is often intercepted by plants. Once this vegetation 'store' is full, rain continuing to fall can reach the surface. Finding out what happens to water beneath the surface is a difficult business. Figure 3.5 shows the amount of seepage from different depths in the soil. The first water to arrive in the stream from the land around is provided by **overland flow**. The graph (B) suggests there is not a lot of this, since infiltration rates of soils are often greater than rainfall intensities. More important is the water moving downslope through the soil. This **throughflow** can feed water into a stream channel fairly rapidly, especially when larger holes form a network of 'pipes' within the soil on a slope. Overland flow and faster throughflow are called *quickflow* processes; they allow rainwater to arrive at the stream soon after the start of a storm.

The other origin of water for the stream lies in the deeper and slower transfer processes. These are known as *delayed flow* processes. The slowest of these, **baseflow**, seeps from the groundwater store. This sustains stream flow between storms.

Figure 3.5 indicates that water arrives at a stream from quick and delayed flows. It reflects the speeds in fig. 2.10. The rainwater, following different pathways, produces the kind of stream behaviour shown earlier in fig. 3.4C. This routing of the water is summarised in fig. 3.6.

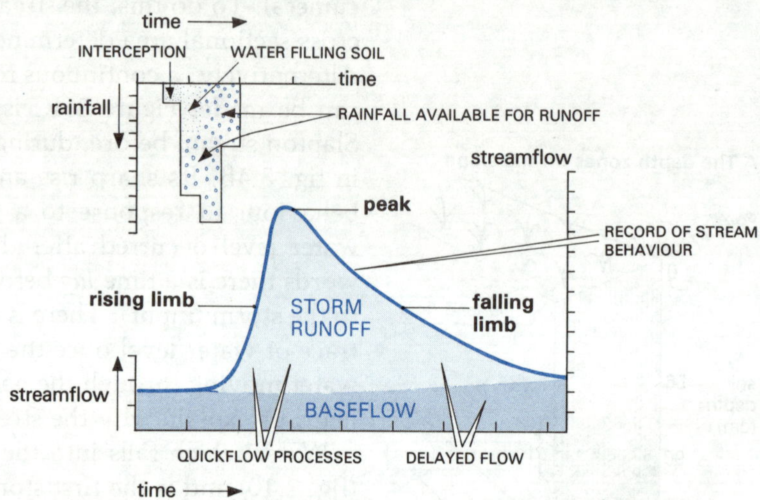

figure 3.6   The flood hydrograph. Water flow over time and a summary of points from the introductory section

A single storm in a larger basin: the River Wye

Understanding how floods are produced and how the flood peak moves like a wave down the river system is basic to flood *prediction*. If we can foresee how big the flood will be, when it will arrive and how long it will last, useful flood warnings can be issued. Even a few hours notice can be crucial. For example, it can give farmers enough time to move stock to higher ground and local authorities time to arrange traffic diversions.

So far we have looked at a tiny basin in Devon. The same kind of short-term stream behaviour can be seen in larger basins. Figure 3.7A shows the basin of the River Wye. It drains an area of 4183 km², stretching from the high plateaus of mid-Wales eastwards to the Bristol Channel. On the 6th August 1974 a storm occurred over part of the basin. Figure 3.7B shows the time and amount of rainfall. Figure 3.7A indicates where this storm's rain fell: the isohyets (lines of equal rainfall) show it concentrated over the mountains in the west of the basin.

28

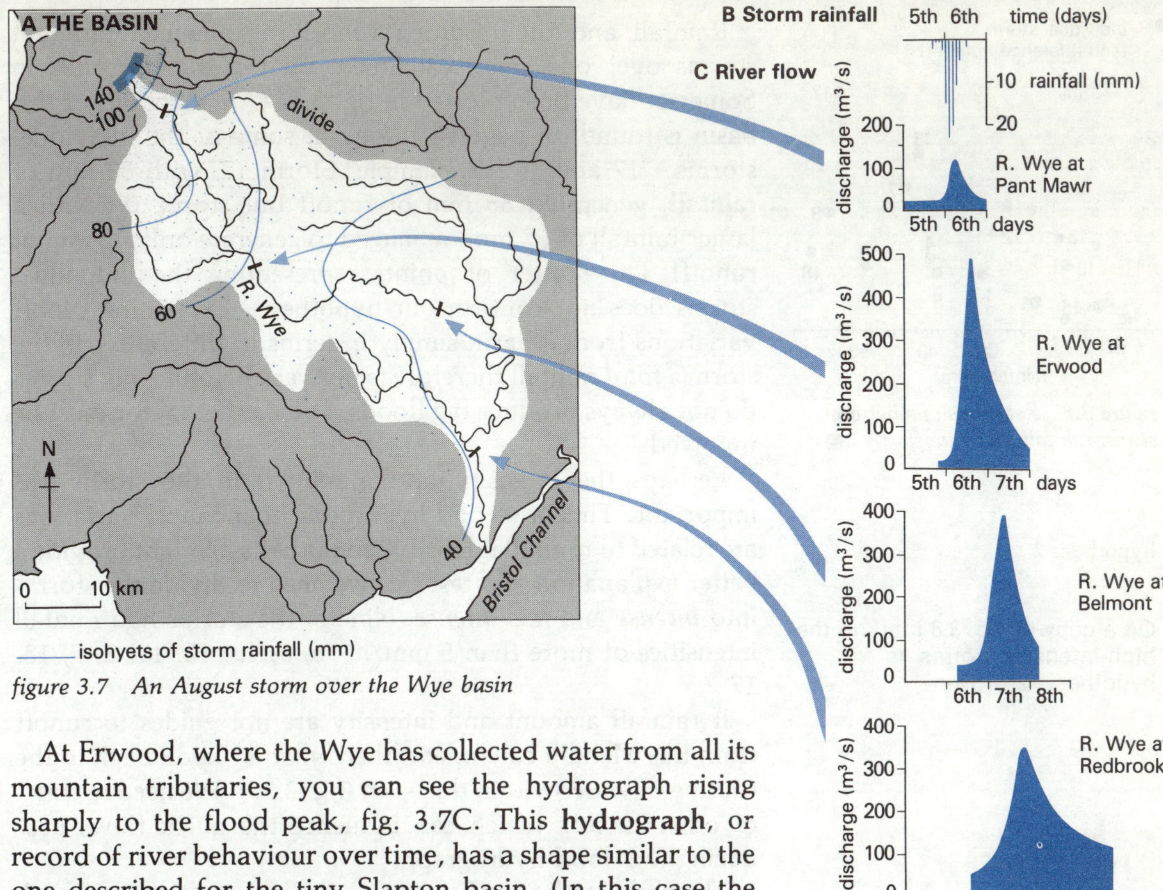

——— isohyets of storm rainfall (mm)

figure 3.7   An August storm over the Wye basin

At Erwood, where the Wye has collected water from all its mountain tributaries, you can see the hydrograph rising sharply to the flood peak, fig. 3.7C. This **hydrograph**, or record of river behaviour over time, has a shape similar to the one described for the tiny Slapton basin. (In this case the quantity of water has been expressed as discharge, in cumecs.) Flood hydrographs are also drawn for Belmont, 68 km downstream from Erwood, and for Redbrook a further 82 km along the river.

The decline in the hydrograph's sharpness and the increasing lag between the storm and the flood peak are typical of the passage of a flood wave down a large river.

How a small basin responds to different storms

So far we have looked at how single storms produce flood hydrographs. We now look at what happens in the basin during and after different storms. The use of a scattergraph to test a hypothesis was introduced in the last chapter (fig. 2.6). This technique can be used to explore the question of why runoff varies after different storms in a basin. We begin with the same hypothesis used then, that there is a positive relationship between rainfall and runoff. In other words 'the more rain that falls in a storm the larger the runoff' (hypothesis 1).

The problem

hypothesis 1

29

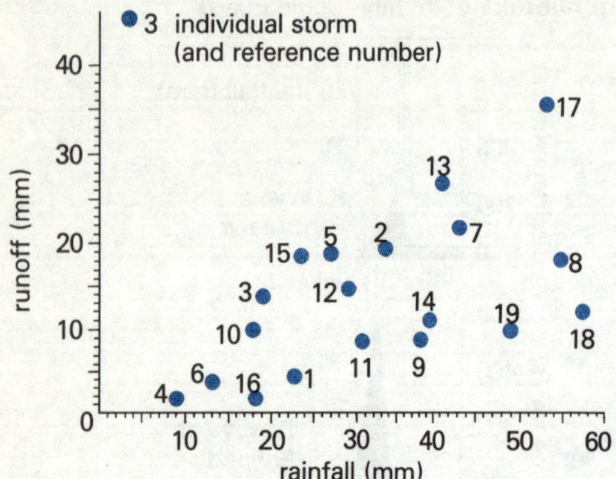

figure 3.8   Relationships between storm rainfall and runoff

hypothesis 2

On a copy of fig. 3.8 mark in the high intensity storms. Is hypothesis 2 proven?

hypothesis 3

1   Emphasise, by using a different colour, the wet storms on a copy of fig. 3.8.

2   Is there a positive relationship under wet and under dry conditions?

3   Why should this be?

Rainfall and the resulting runoff for nineteen different storms over one small catchment on the Mendip Hills in Somerset have been plotted in fig. 3.8. (An air photo of this basin is found on page 61.) Look at some of the individual storms – 17 and 18 for example. Storm 17, with 53 mm of rainfall, generated 36 mm of runoff but storm 18, with a larger rainfall of 57 mm, managed to generate only 12 mm of runoff! The scatter of points representing the individual storms does not confirm our hypothesis. Explaining runoff variations from a basin simply in terms of differences in the storm's *total* rainfall therefore is not very helpful – big storms do not always produce big floods. Some other factor must be involved.

Perhaps the intensity of the rainfall in the storm was important. Thus a second hypothesis 'that runoff variations are related to rainfall intensity' (hypothesis 2) might provide a better explanation. To test this we need to divide the storms into *intense* and *less intense*. Nine of the storms had rainfall intensities of more than 5 mm/hr: 6, 3, 15, 12, 11, 2, 7, 18, 17.

If rainfall amount and intensity are not guides to runoff from this Mendip catchment, then what is? Look back to the routing of water within the basin (fig. 2.10). We need to think of other factors which can influence the quick flows producing the flood hydrograph.

The characteristics of the ground surface affect how much water goes into or over the soil. From what was said in the last chapter you might have begun to think that vegetation and soil texture could be factors. In the case of our Mendip catchment they are unchanged – the land use has not been altered and the soil texture cannot change between storms. One thing which can alter is the *amount of water already in the soil* before the storm begins. If the soil is saturated, incoming rain will have to move overland and relatively quickly to the stream channel. This idea could be turned into our third hypothesis – 'that runoff will increase with rainfall and soil moisture' (hypothesis 3).

To test this we need to divide the storms into two sets, those when soil was 'dry' and those when it was 'wet'. The 'wet' storms (when more than 12 mm of rain had fallen in the five day period before the storm) were 4, 6, 10, 3, 12, 15, 5, 2, 7, 13, 17.

Whether a soil is 'wet' or 'dry' reflects what can be termed the **antecedent moisture conditions**. Your plot, therefore, illustrates the fact that these conditions influence storm runoff.

A dynamic model of the basin

How a small catchment behaves under 'dry' or 'wet' conditions is summarised in fig. 3.9. The **water table** is the upper surface of the zone of bedrock and soil whose pores and fissures are saturated with water. As it rises, the *saturated zones*, at the base of slopes close to the channel, increase. By the end of a wet period, such saturated waterlogged areas could expand to as much as 30 per cent of the catchment's area! Unable to absorb any more rainwater these areas feed water to the streams very quickly, contributing to the rapid rise of the flood hydrograph.

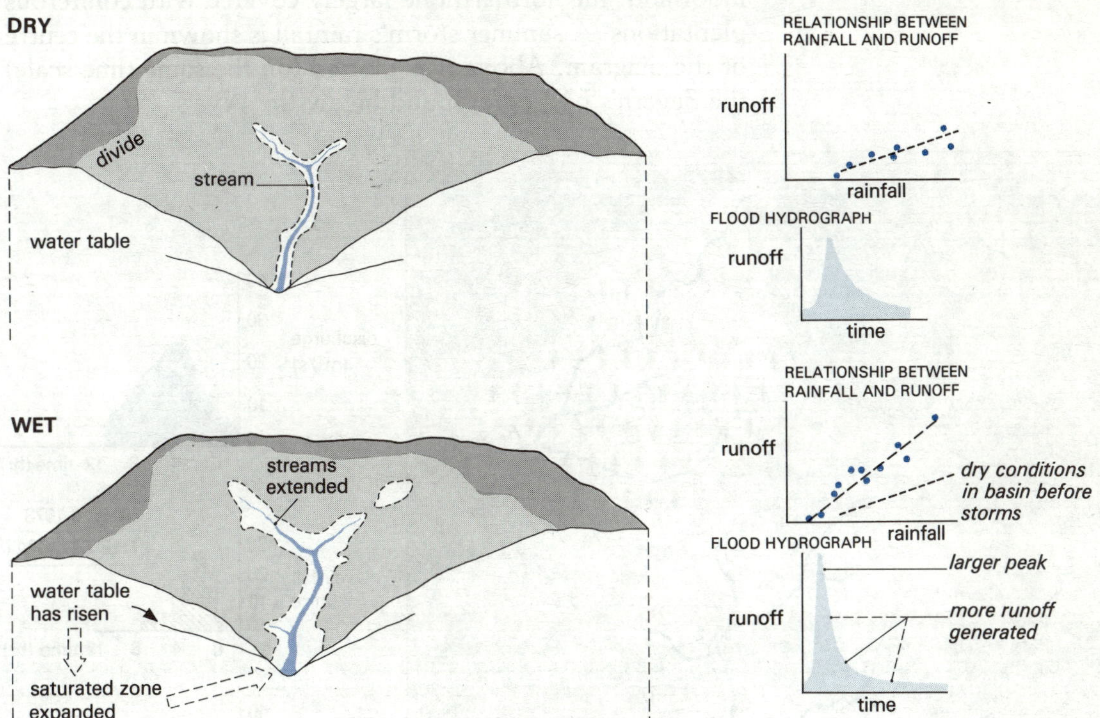

figure 3.9   *Changes in a small catchment under wet and dry conditions*

Water moving downslope within the soil as throughflow arrives at the stream later than the water flowing off saturated sites. After a storm slopes may take more than forty days to drain their water to the stream! Such delayed flow produces the 'falling limb' of the flood hydrograph.

The diagram also suggests that the shapes of the flood hydrographs are different – not only is the amount of runoff greater under wet conditions, the time it peaks is also different.

Look at fig. 3.9 and describe the differences in runoff under wet and dry conditions.

The influence of vegetation and land use on flood hydrographs

How a river responds to storm rainfall can also be influenced by the type of land surface. One way to discover the influence of vegetation is to look at two catchments with *different* plant cover. They need to be *similar* in other respects such as size, slopes, soil, drainage density and so on. We also need to look at how they behave with the same rainfall input.

The Wye and Severn area gives us an example and fig. 3.10 shows two small catchments – the southern one covered with moorland, the northern one largely covered with coniferous plantations. A summer storm's rainfall is shown in the centre of the diagram. Above it is plotted (on the same time scale) the Severn's hydrograph and below the Wye's.

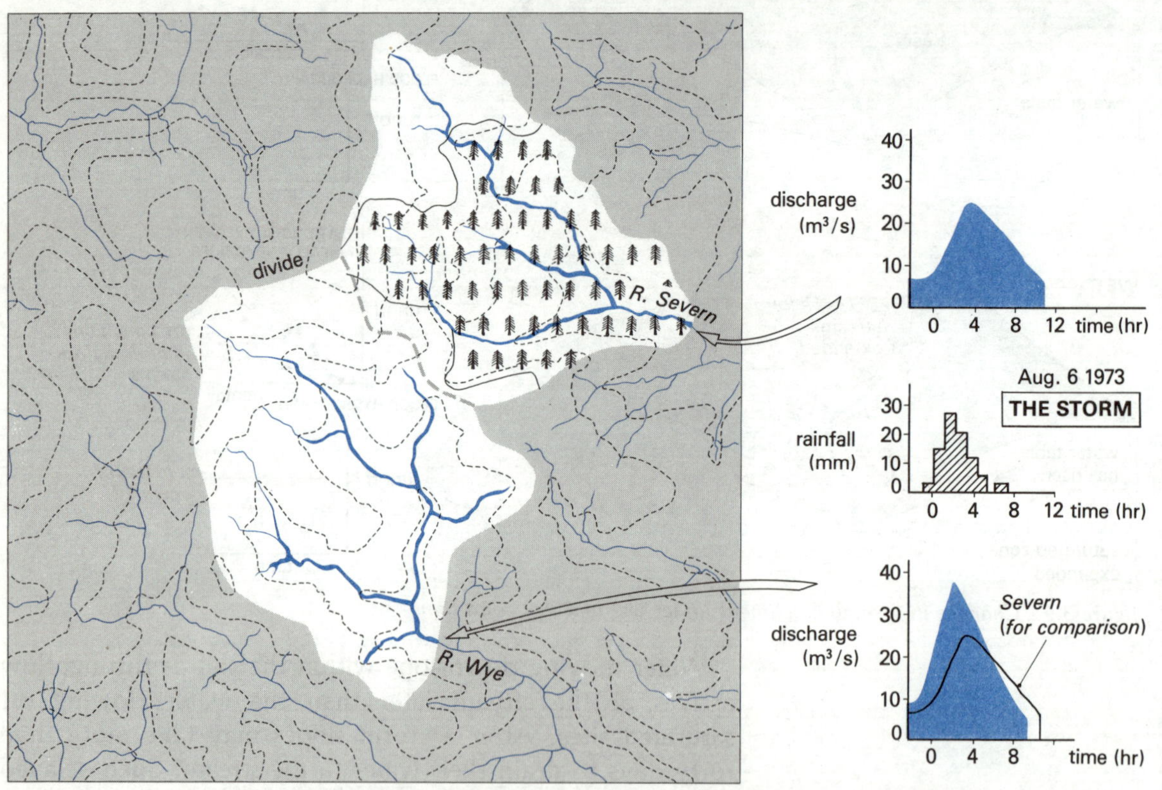

figure 3.10   The upper Wye and Severn

1   Describe the amount and timing of the rainstorm.

2   Describe the shape of the Severn's hydrograph. When did the flood peak occur and how did this compare with the rainfall?

3   Look at the Wye's hydrograph. Compare and contrast its shape with the Severn's.

Seventy per cent of the storm's rainfall ended up as runoff in the forested catchment of the Severn. In contrast, the figure for the moorland Wye was 82 per cent. The Wye had a 'flashy' response to the storm; as you have noticed, its lag time was less, its peak greater and the falling limb steeper than the more delayed response of the Severn.

Man frequently changes the land surface; the Forestry Commission after all had planted the trees in our last example! Man's most dramatic alteration is when he replaces cropland, hedges and trees with buildings, roads and gardens. An example of the effects of urbanisation on stream behaviour comes from Harlow. Most of the New Town is drained by Canon's Brook. Figure 3.13A shows the catchment still covered in farmland and the graph below it shows the stream's flood behaviour. Figure 3.13B shows the situation after the New Town was built. Such changes are typical of those produced by urbanisation.

figure 3.11   A Forestry Commission plantation

figure 3.12   Urbanisation: a New Town, Harlow, covering farmland

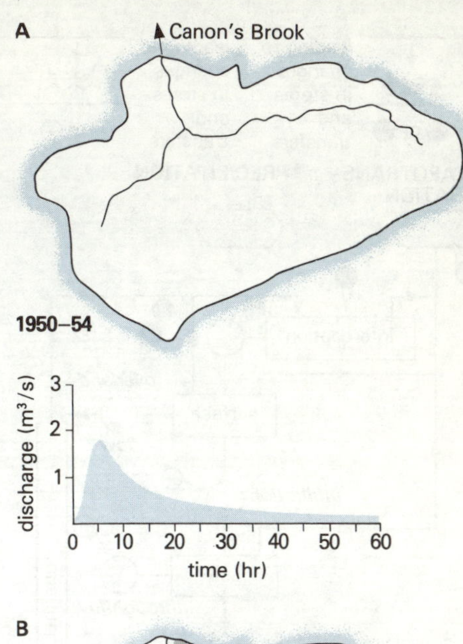

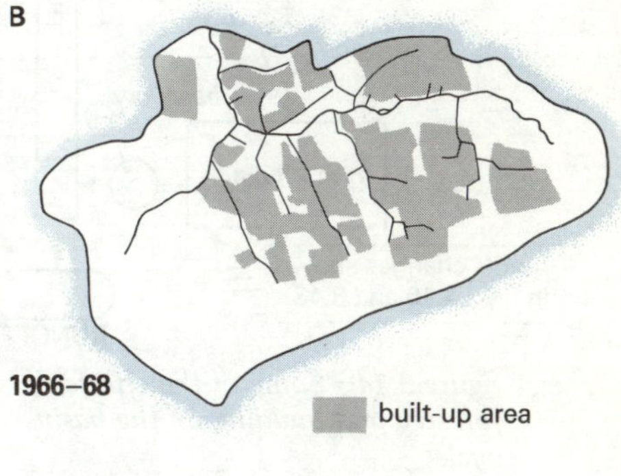

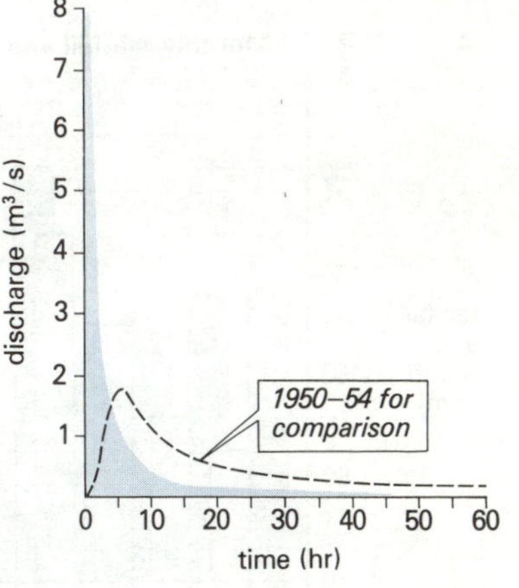

figure 3.13   Urbanisation and stream behaviour, Harlow New Town

Describe the changes in the flood peak, lag and baseflow which occur during the urbanisation of a catchment (fig. 3.13).

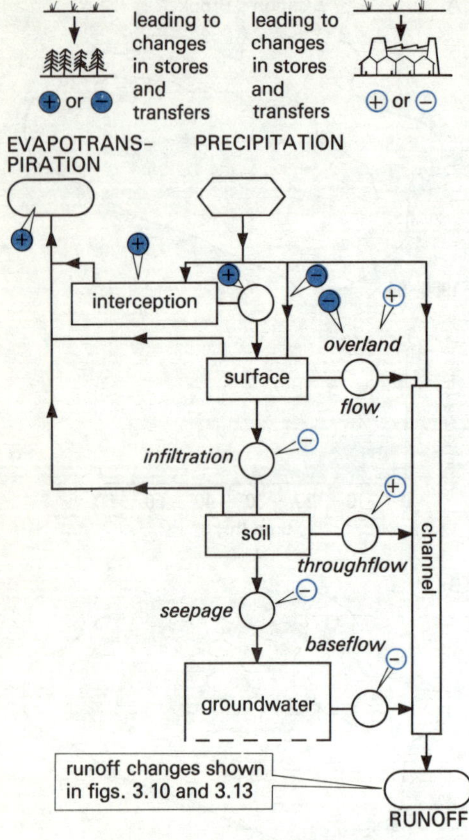

figure 3.14   Some of the effects of forestry and building on the basin cycle

Figure 3.14 summarises some of the effects of forestry and building on the basin system. The base for the diagram is the diagram built up in the last chapter showing the transfer and storage of water (fig. 2.10). The + and − symbols refer to changes in the system which altering land use produces.

Using fig. 3.14 to help organise your ideas, write an account of how man can interfere with a basin's flood behaviour.

Expanding the time scales: the drainage basin over a year and annual water budgets

The water balance

Earlier we looked at a single rainstorm in the Slapton Wood catchment in Devon. Our emphasis now shifts to rainfall and runoff over the longer period of a year. The Field Centre also keeps such longer term records. Figure 3.15 shows rainfall and runoff for October 1976 until September 1977. The differences between the rainfall and runoff amounts represent the water 'losses'.

Of course the water is not really lost, it is cycling through pathways which do not involve stream runoff. High summer temperatures provoke increased evaporation but the major difference between summer and winter is due to plant growth. Plant roots tap the water received and stored in the soil and their leaves transpire it to the atmosphere during the growing season.

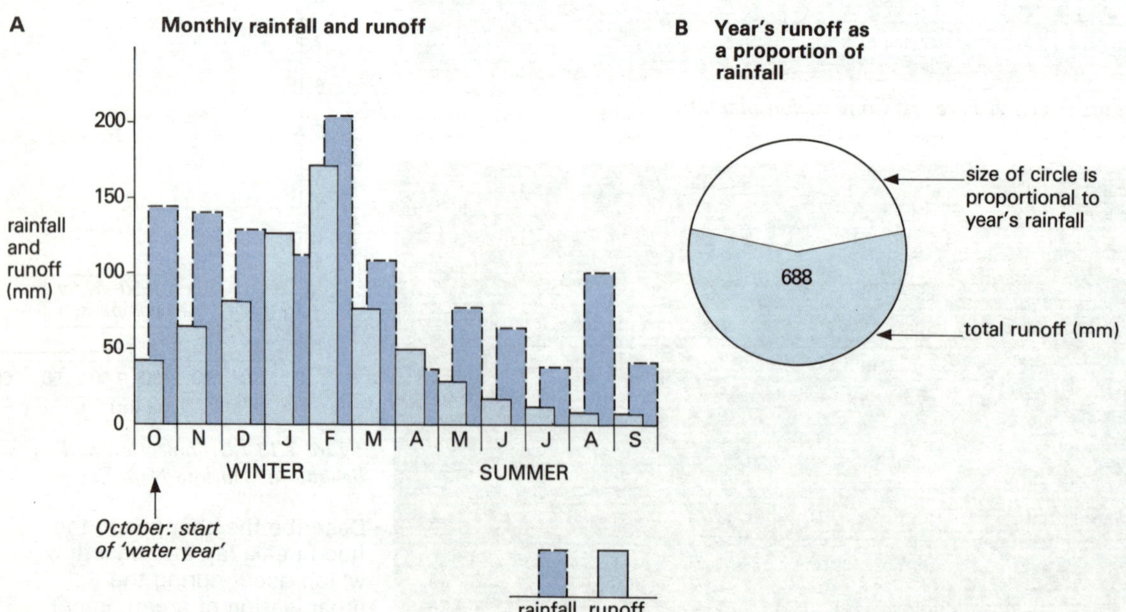

figure 3.15   Slapton Wood catchment: a year's rainfall and runoff

There were two months in this particular year when runoff exceeded rainfall, January and April. This is not as bizarre as it sounds, as measurements are totalled for calendar months. A storm on December 30th, for example, would have its rainfall included in the December total. Its runoff, particularly the delayed flow, would be included in January's total.

Figure 3.15B is a circle whose area represents the year's rainfall over this catchment(1211 mm). The shaded sector shows the runoff proportion (688 mm). The unshaded sector is evapotranspiration. Another way to look at this is to imagine a yearly balance (or budget) of water where the input balances the outputs from the basin. This can be written as:

rainfall = runoff + evapotranspiration ± storage changes.

Storage changes is included for completeness, but what does it involve? Imagine a basin underlain by porous rocks. In a wet year, water added to the groundwater store produces a rise in the water table. In dry years, slow seepage from the rocks, pores and fissures occurs to maintain streamflow and the water table falls.

Where man uses groundwater, changes in the groundwater store can be detected as water table changes in wells and bore holes. However, in most cases it is difficult to measure. It is for this reason that in Britain the 'water year' starts in October. After the summer's evapotranspiration the groundwater store is at its smallest, so any inaccuracies in expressing the year's water balance are reduced.

Variations in the water balance

Areas with different climates have different water balances. An example of the British situation can be seen in fig. 3.16 which shows a year's monthly rainfalls and runoffs for two Scottish basins. The Leven in the west drains high mountains and Rannoch Moor. In the lowland east the Esk drains into the Firth of Forth. Only 150 km apart, the basins show marked differences in monthly rainfall and runoff as well as in their annual balances (shown by the pie graphs).

1 Describe the monthly rainfalls and runoffs. What differences are there?

2 What differences are there between the basins' annual water balances?

The upland western basin, the Leven, not only has more rainfall, it has higher runoff in total *and* also as a percentage of rainfall. In Britain the predominant airflow is from the west. Depressions moving in from the Atlantic meet the

1 Look at fig. 3.15 and describe the pattern of rainfall and runoff for the year 1976–77.

2 Why does our 'water year' run from October?

Water balance

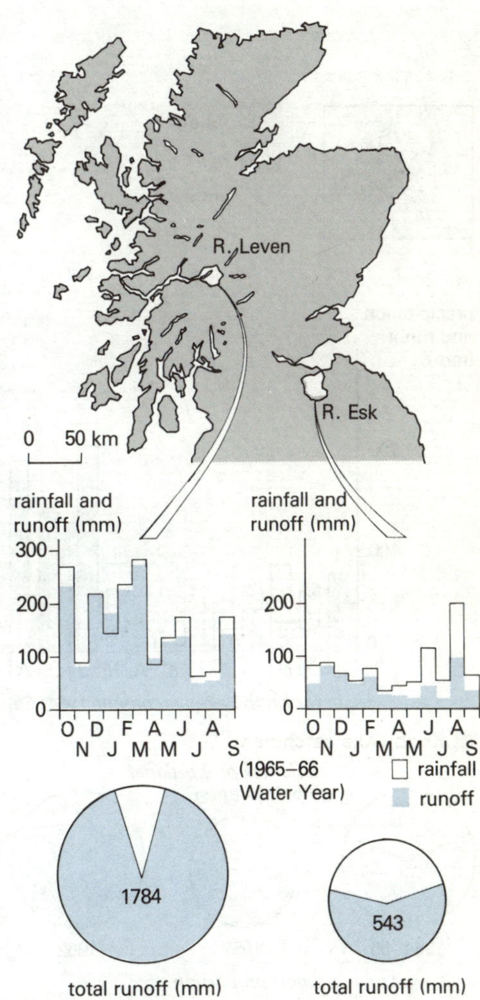

figure 3.16 A year's precipitation and runoff for two Scottish basins

mountains. Frontal rainfall connected with the depressions is added to by a relief component as the airstream is forced upwards. Such uplift causes cooling and more precipitation over the mountains. The higher and cooler western moorlands, like those of the Leven basin, also have lower evapotranspiration losses. For Britain the supply of water (runoff) is greatest in the higher western areas. Most of our cities, people and industries using water however are in the lowlands and the east. This mismatch between the location of supply and demand has an important influence on water resource planning in Britain.

1 For your local area discover and describe: (a) the source(s) of its water supply, (b) the seasonal pattern of demand, (c) where within the area the water is used (e.g. domestically, industrially, agriculturally) and (d) any plans for improvements to water supply and quality.

2 Using atlases and books research the UK's water resources. Produce a series of maps to summarise supply, demand and how water is moved to areas of demand.

Fig. 3.17 shows a year's monthly precipitation and runoff for the sub-arctic Knob Lake catchment in Canada. Notice a mere 'trickle' of water, less than 0.25 mm a day, is running off beneath the snow and ice during the winter. The winter snowpack 'locks up' the precipitated water in a surface store

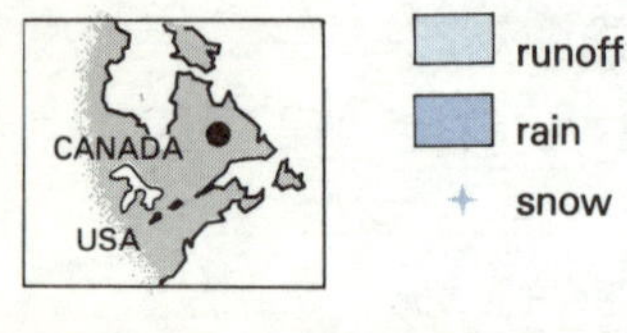

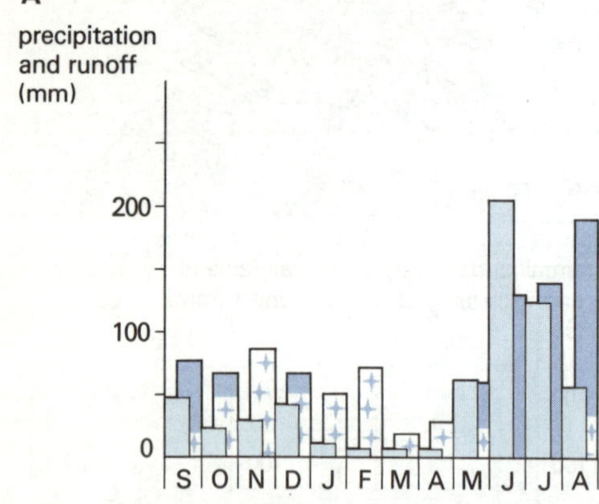

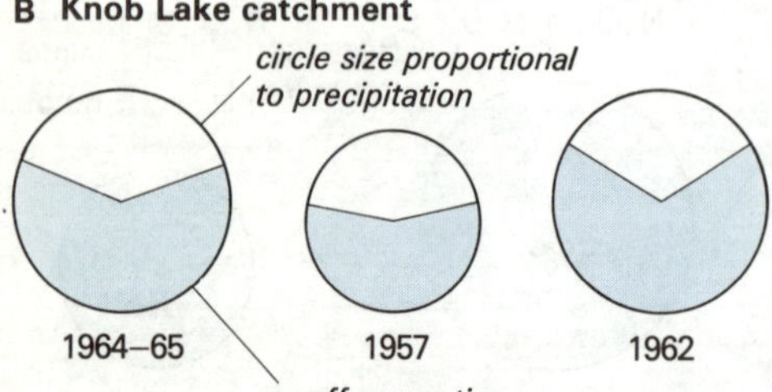

figure 3.17   Precipitation and runoff in a sub-arctic catchment

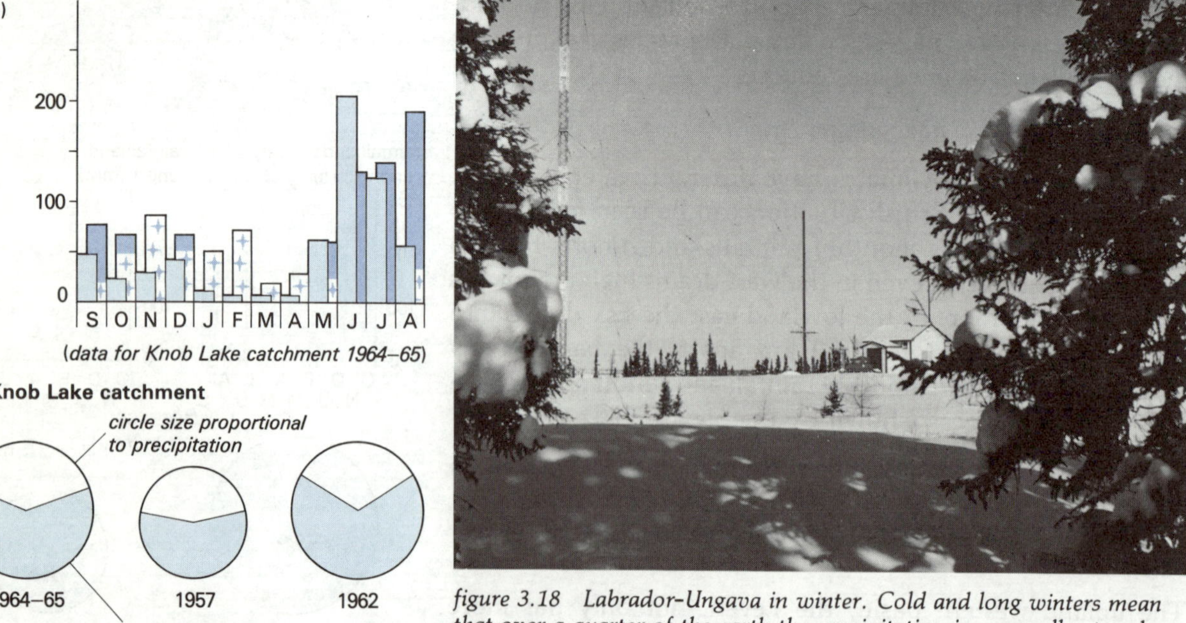

figure 3.18   *Labrador-Ungava in winter. Cold and long winters mean that over a quarter of the earth the precipitation is seasonally stored up as a snowpack. When this melts in spring it produces a flush of water to the streams. Lakes help to even this out, by storing some of the spring's snowmelt and releasing it during the summer.*

from October to May. The runoff of 0.86 mm on May 17 had risen to 6.4 mm by the end of the month. With its low winter runoff and rapid rise to an early summer peak after snowmelt, this catchment is an example of a snowy environment.

1 Using fig. 3.17, make a calendar showing how much water arrived (as rain or snow) and how much left as runoff each month.

2 If you had to develop the hydro-electric power potential of this kind of area what problems would you have to try and solve?

Runoff variations from year to year

Individual water years were the focus of the last section. Wet years, dry years, warm years, cold years and the whole range of climatic possibilities affect an area's runoff. Figure 3.17B shows annual precipitation and runoff for our sub-Arctic catchment. Data for two years has been plotted, one slightly 'wetter' and one 'drier' than the year whose monthly figures were looked at earlier (fig. 3.17A).

Such variations from year to year are found in Britain too. Twenty-three years' records for the River Wye at Erwood (see fig. 3.7) are plotted in fig. 3.19. Each month's runoff is shown by a dot in the monthly column and the 33-year average is shown by the solid line. This average line reflects the familiar British situation.

The graph illustrates the variation in runoff for particular months. The scatter of points also indicates that the range of runoff between the highest and lowest figures is greater in some months than others. For example compare February and July. The dots representing the months of one particular year have been emphasised. You can see that in some months runoff for this particular year was above the mean, whereas in others it was below it.

In Britain, climatic variations are not normally extreme enough to have a *widespread* impact on society. River flooding affecting parts of towns and areas of farmland of course occurs more frequently. The 1975–76 drought was an exception. This sixteen months was the longest dry period in the United Kingdom since records began in 1727.

Some effects of the 1976 drought are shown in fig. 3.20. The distribution of reduced runoff and streamflow is shown in fig. 3.20A. The map indicates the 1975–76 runoff as a percentage of the long-term mean: the darker the shading the greater the reduction in streamflow. The discharge of two rivers, the Dee in the north-west and the Chelmer in the east is also shown.

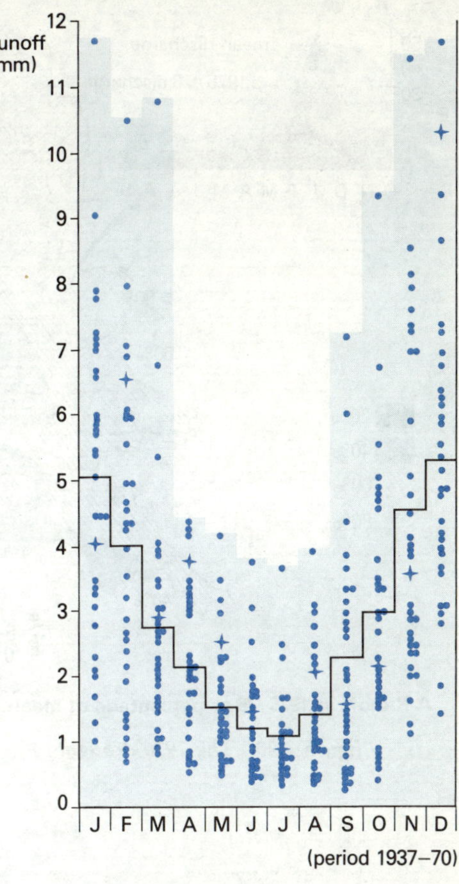

figure 3.19   *Variations in monthly runoff for the River Wye at Erwood*

Look at fig. 3.19 which plots 33 years' records.

1 Which months have the greatest and smallest ranges of runoff? What implications might this have?

2 Describe and explain the average runoff.

3 How did the one year plotted (1965–66) compare with the long-term situation?

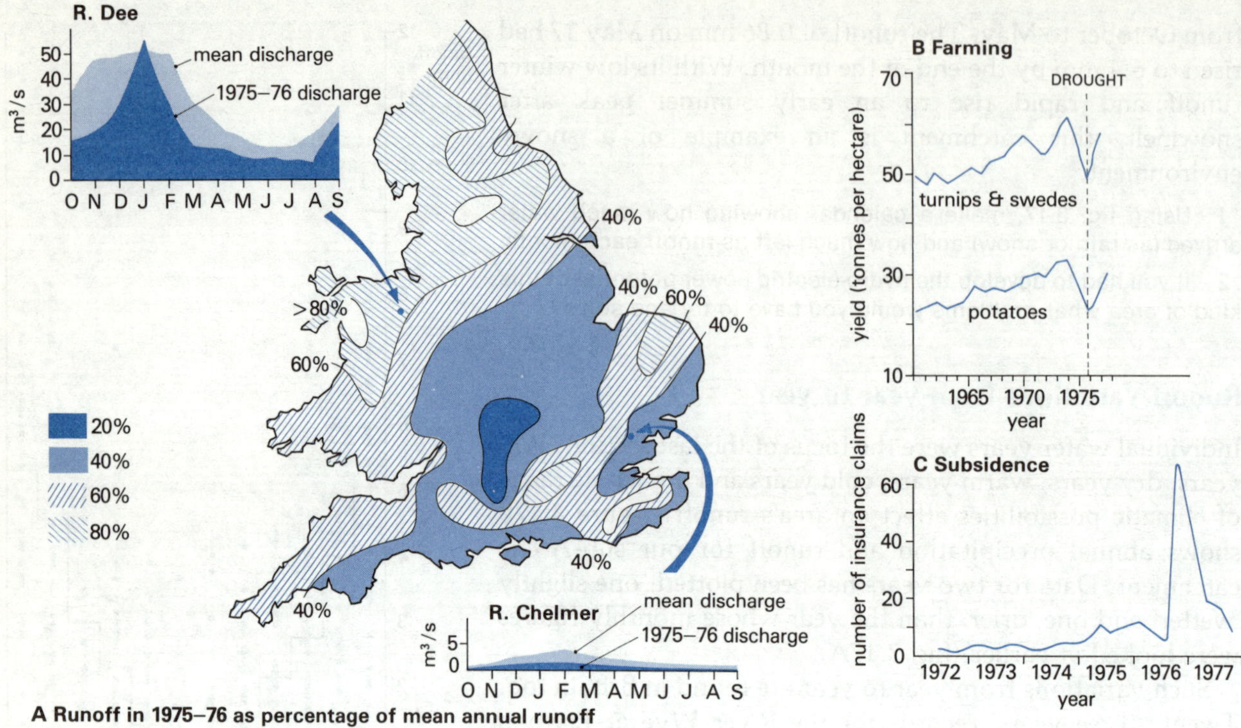

A Runoff in 1975–76 as percentage of mean annual runoff

figure 3.20   The 1976 drought

The drought had a number of impacts. Farming was affected in many ways, such as reduced grazing and the lower yields of some crops (fig. 3.20B). The sharp dip in the graphs reflects the reduction in the amount of water available in the soil during the drought (a smaller groundwater and soil store in terms of fig. 2.10). The removal of water from the soil also had the effect of 'shrinking' many clay soils. You may have seen this shrinkage of clay in the cracks which form during normal summer dry spells. During the drought, however, this shrinkage was more severe and many buildings built on clay suffered structural damage as their foundations moved. An example is shown in fig. 3.20C which shows insurance claims rising dramatically towards the end of the drought.

In many parts of the world climatic variations are more critical. One example is from India where the arrival of the **monsoon** and its summer rainfall are a dominant part of the physical environment with important human consequences. Table 3.1 is a table of monthly average discharges for the River Tapi. If you look at 1959 you can see that the highest monthly discharge was 614 times the lowest! The variations from high flows, once the monsoon rains have started, to the lower flows are so large that they stand out from the figures without the need for a graph like fig. 3.19.

38

table 3.1   River Variations in a Monsoon climate: River Tapi, India.

YEAR	MONTH											
	Jan	Feb	Mar	Apr	May	Jun	July	Aug	Sep	Oct	Nov	Dec
1950	110	84	67	47	28	39	647	851	1474	291	81	55
1951	52	35	27	21	17	199	788	1061	339	336	49	31
1952	24	20	14	10	8	205	571	781	167	91	19	13
1953	13	11	9	6	5	389	662	2331	945	284	52	23
1954					no records available							
1955												
1956						1282	2297	2346	1302	1321	622	303
1957	101	137	116	97	62	486	653	1218	638	238	87	50
1958	41	41	22	17	22	78	1621	2195	2393	510	123	154
1959	21	84	78	11	21	88	1869	2103	6755	980	140	140
1960	70	42	19	21	26	184	310	296	203	134	36	211

The figures are monthly average discharges in m³/s
The basin area is 61 575 km²
Blue figures show the highest monthly average in each year

If we consider the largest drainage basins, they may drain areas with different climates. The behaviour of the main river near its mouth could then reflect a combination of various climates in different parts of the basin (fig. 3.21A). Figure 3.21B shows variations in output from the Mississippi River over thirty-six years. Each dot is the average discharge for a particular year. The map shows the mosaic of flood seasons within the basin's 3200 km².

A Flood seasons

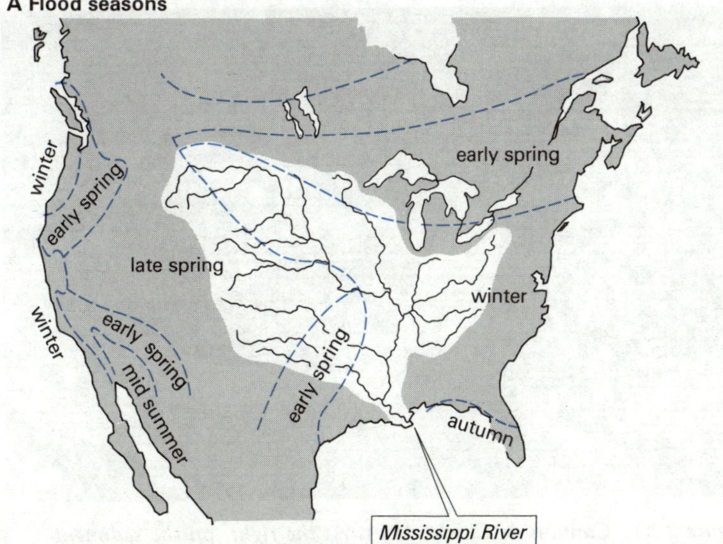

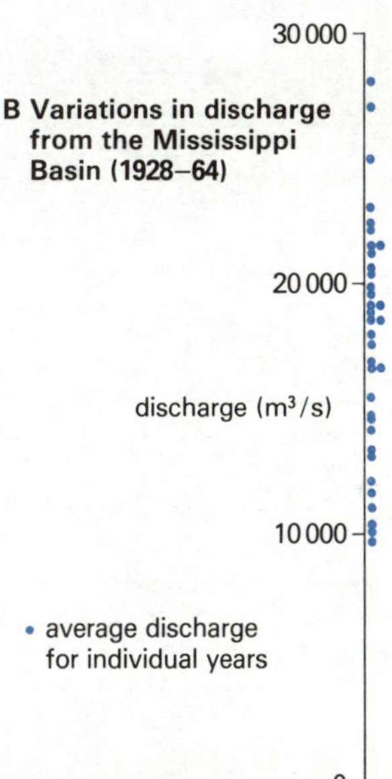

figure 3.21   The Mississippi basin: flood seasons and discharge variations

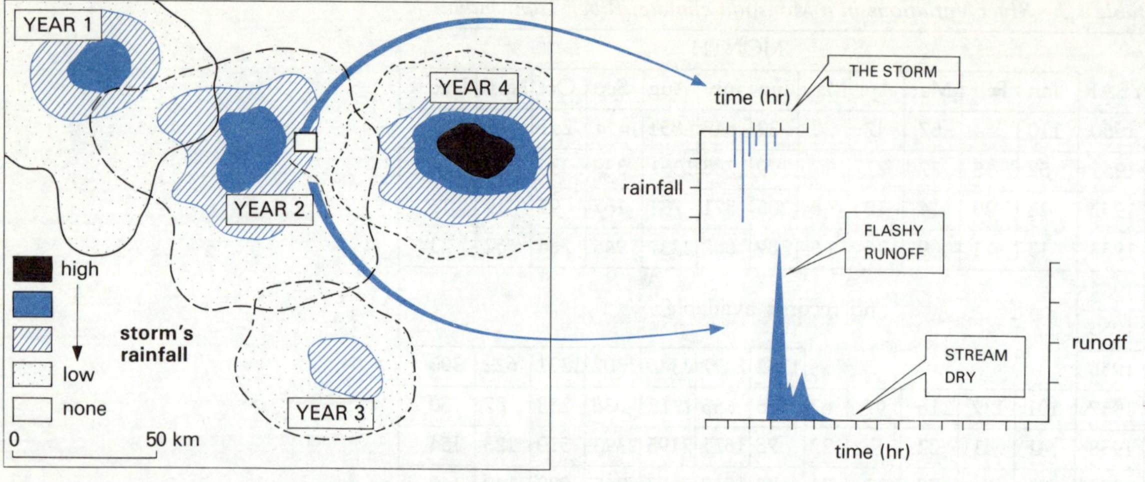

figure 3.22   *A model of arid zone storms and runoff*

So far we have not mentioned the third of the earth's land surface which experiences desert or **arid** conditions. In these areas average rainfall totals are low and plant cover sparse. Stream behaviour here is very different from examples discussed earlier. Figure 3.22 summarises the situation. As the diagram shows, rainfall is scattered or patchy, with storms providing rain for distinct areas. The storms may also be occasional or spasmodic in timing, with months and years (or even decades) between them. When sufficient rain falls to generate runoff the stream's response is flashy. The dry channel beds are rapidly occupied by water and there is a quick rise to a flood peak (fig. 3.22).

figure 3.23   *Canyon de Chelly, USA. At the right, on the sediment-strewn valley floor, can be seen the last dregs of streamflow from a recent storm.*

Conclusion

Before leaving the topic of runoff variations, it is interesting to consider what happens over a time span of thousands of years.

If time travel was possible, what kind of earth might you have seen from a spacecraft 20 000 years ago? Compared to earth today there would be major differences in water distribution. If you looked at America and Europe you would have seen glistening, massive spreads of snow and ice. The North American ice sheet was more than 5 km thick. Locking up water in such huge stores meant a world-wide fall in sea level of almost 200 metres. This exposed areas which are now shallow seas as land, so you would find it difficult to recognise coastal shapes. Interlinked with the ice sheets was a wholesale shifting of climatic regions. Parts of Europe not covered by ice would have experienced tundra climates where river behaviour would have been dominated by summer snowmelt (rather like fig. 3.17). Further south, areas now desert would have been better watered and Stone Age hunters in the central Sahara created cave paintings showing the types of animals found in savanna grasslands today.

Rivers which behave within certain limits today, in the relatively recent geological past may have had to respond to very different rainfall and runoff situations. As the direction of the earth's climatic changes is still not understood, the Ice Age scene described above might also represent the world of our descendants, in less than 200 generations!

Summary

- The chapter examined runoff in small and large basins. It also looked at river behaviour over periods of a few hours and over years. It therefore expanded the ideas of chapter 2 by looking at variations in the drainage basin hydrological cycle.
- The flood hydrograph is produced by quickflow processes.
- Flood behaviour depends on soil (especially moisture), and on land use as well as on weather and climate.
- The water balance (or budget) is a key idea when examining a basin's rainfall and runoff.
- Water balances vary in time and space.
- River flow is a resource and a hazard (i.e. when there is too much or too little water).

1 How do you think the catchment shown in fig. 3.4 would behave if the following happened (you could sketch the shape of the flood hydrograph and give notes for your reasons):
(a) the catchment had been wetter before the storm;
(b) all the land surface had been covered with coniferous trees;
(c) a housing estate was built over half of the catchment.

2 Data listed in table 3.2 is from a small stream on Long Island, New York.
(a) Plot the data in table A on a scattergraph. Is there any relationship between rainfall and runoff?
(b) On the same graph, but using a different colour, plot the data for a later twelve-year period (part B). Describe the relationship between basin input and output.
(c) Describe the differences between A and B.
(d) Data B is for the period after 1952. What could have happened after 1952 to produce this?

3 Table 3.3 shows average rainfall and runoff (in mm) for ten British catchments.
(a) On a base map of England and Wales plot the locations of the basins and draw proportional circles like those in fig. 3.16 (or divided bar graphs) for the basin's rainfall and runoff.
(b) Where does most rainfall occur?
(c) In which parts of the country is runoff lowest?
(d) Explain the patterns of rainfall and runoff.
(e) Briefly discuss what Britain's water resource strategy could be.

table 3.2 Annual rainfall and runoff from a Long Island catchment

	R	R/off	R	R/off
A	1450	91	900	55
	1240	112	1090	73
	1350	150	990	85
	1480	117	1140	81
B	820	24	980	29
	860	31	990	32
	925	34	1190	42
	1470	41	1170	45
	1150	37	1270	59
	1000	45	1210	52

R = precipitation (mm)
R/off = runoff (litres/sec)

table 3.3

Basin	R	R/off	Location & National Grid reference	
Greta	2270	2204	Cumberland	NY 309191
Wye	2326	1806	[see Fig. 3.7] Pant Mawr	SN 843825
Chew	986	458	Avon	ST 648647
Tamar	1252	792	Cornwall	SX 426725
Piddle	918	412	Dorset	SY 913876
Darwell	950	325	Kent	TQ 722213
Mimram	657	125	Essex	TL 282133
Bure	699	208	Norfolk	TG 192296
Witham	635	164	Lincoln	SK 842480
Gaunless	751	354	Durham	NZ 215306

R = precipitation (mm)
R/off = runoff (mm)

4 Decay, destruction and movement on the earth's surface

An Italian disaster

In the autumn of 1960 engineers proudly viewed the completed Vaiont Dam, a beautiful slender concrete arch spanning the Vaiont gorge in northern Italy. Soaring 265 metres, it was the second highest dam in the world.

Three years later, at 10.45 p.m. on October 9th, disaster struck. A section of hillside 1800 m long, 1600 m wide and 150 m thick suddenly moved down the valley side and plunged into the reservoir. The impact of this huge mass of rock threw water 260 m up the opposite slope. The dam held, although it was overtopped and a 'wall of water' 70 m high rushed down into the Piave valley and the town of Langarone. A total of 2600 people were killed and Vaiont entered the record book of major disasters. Tracing the reasons for this disaster introduces the chapter themes: rock breakdown and movement.

Why did the disaster occur?

In explaining the Vaiont disaster we need to look at its *setting*. The steep-sided gorge, which provided such an 'obvious' site for the dam, lay within a larger valley. The rock layers underlying the valley had been warped into a downfold so that they dipped towards the centre of the valley. The cross section of the valley (fig. 4.1) shows this tilting of the rocks, which had been produced by earth movements.

The inner gorge, the dam site, was cut into limestone. Above this the layers of limestones were interbedded with clays. These limestones were riddled with fissures and cracks, allowing rainwater to pass into the rock, which meant that water could easily reach the clay. Those of you who have slithered on the ground in matches played on wet clay pitches might appreciate how tricky the situation could become.

Whether rock material stays where it is, or moves downslope, depends upon the *balance of forces*. Trying to move material downslope are the weight of the rock, the steepness of the slope, the pressures exerted from further up the slope and the force of gravity. When these downslope

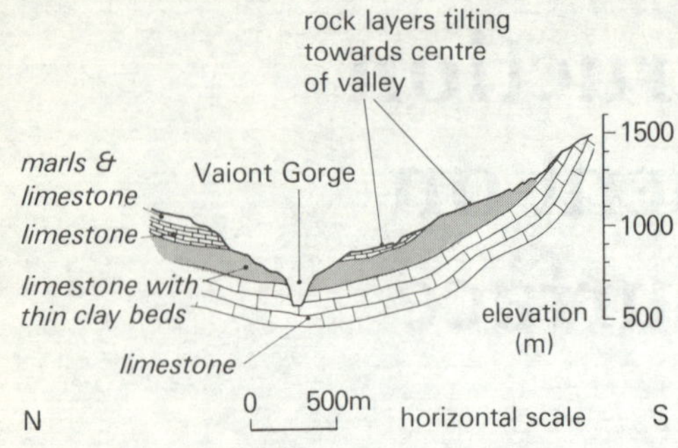

figure 4.1   Vaiont, the basic situation. A cross section of the valley showing rock strata.

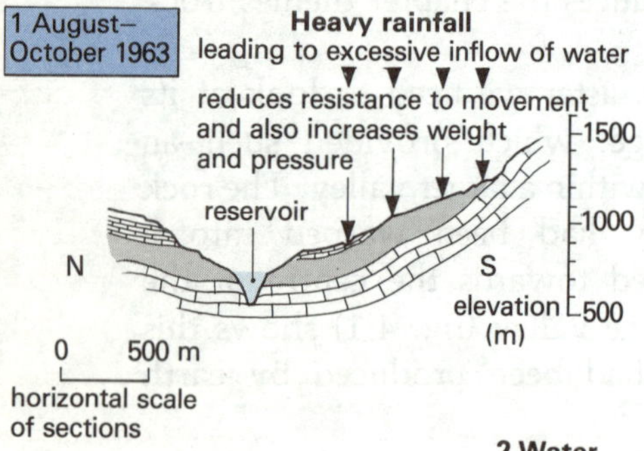

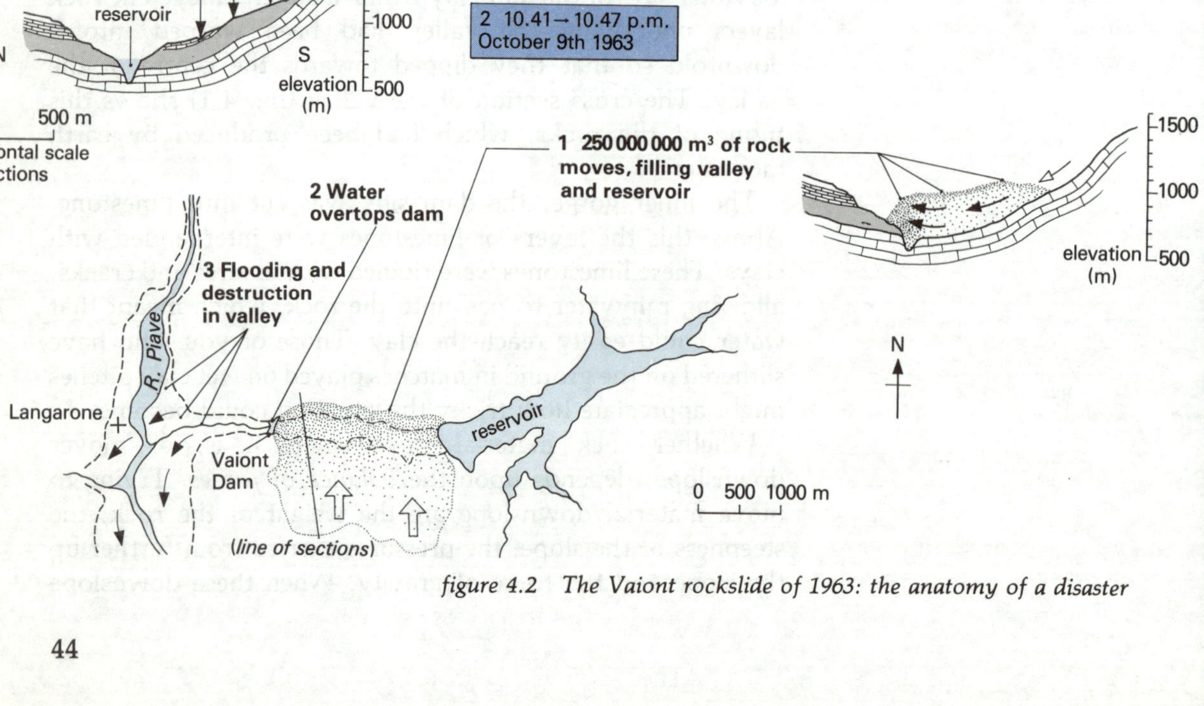

figure 4.2   The Vaiont rockslide of 1963: the anatomy of a disaster

forces are weaker than the resistance (friction) opposing them the rock and debris stays where it is. When the downslope forces are greater then the rock moves.

In May 1963, six months before the disaster, engineers measured downslope movements of up to 1 cm a week, in geological terms a quick process. After the heavy rainfalls began on September 28th they measured movements of up to 40 cm a day! By October 1st so unstable had the hillslope become that wild animals, apparently sensing the danger, stopped grazing on the slope. *Something was upsetting the balance between downslope forces and friction.* Anxious engineers began lowering water levels in the reservoir on October 8th – but too late: the massive rock slide occurred the next day. What caused the 'slope failure' which dumped a quarter of a million cubic metres of rock into the reservoir?

The reasons for this sudden movement are summarised in fig. 4.2. The heavy rainfall had two effects. Firstly, the water falling onto the soil and filling the fissures, cracks and pores in the rock increased the weight of material on the slope. As if this was not enough the rain had a second effect. Within the clay bands this water increased the 'fluid pressure' in the pores between the tiny clay particles. This reduced their resistance to downslope forces. So the rain upset the balance by making the downslope forces greater as a result of increasing weight. At the same time it made the other side of the scales lighter by reducing friction. Upsetting the balance of forces in these two ways caused the sudden rock slide of October 9th.

What happened at Vaiont was a result of a combination of factors: some related to the kinds of rock (limestones fissured by cracks), some to the ways the rocks were arranged (tilting or dipping downslope with layers of clay between the beds of limestone) and some to the weather conditions (the heavy rainfalls before the slide). Before we look at other examples of how rock materials move we need to define 'rock' and discover how it is altered.

Imagine you were a reporter covering the Vaiont disaster. Your editor has allowed you two short paragraphs, a simple diagram and one photo. What would you write, draw and photograph to explain the disaster to the general public?

Rock classification

Tiny atoms make up all substances, whether they are solid, liquid or gas. Atoms of the same substance bonded together form elements. Some of these are found in the earth's crust, copper, gold and sulphur for example. It is more common, however, to find different elements joined together as compounds. Elements and compounds found in the earth are known as *minerals*. There are about two thousand different minerals, but only about thirty are common. Rocks can be divided into three types according to their formation – igneous, sedimentary and metamorphic.

Igneous rocks formed from molten rock. As magma cools minerals change from their liquid to solid states. Magma erupted onto the earth's surface cools quickly so mineral crystals have no time to grow and the resulting rock is fine grained, like the basalt mentioned in chapter 1, for example. When magma cools slowly larger crystals can form, like the mica, quartz and feldspar crystals in granite formed below the surface. Igneous rocks, therefore, can have fine or coarse textures. Another subdivision reflects the amount of silica in the rock: when this is more than 65 per cent the rocks are *acid*, when less than 55 per cent the rocks are termed *basic*. Igneous rocks, therefore, may be classified according to their *texture* and their *chemical composition*.

Sedimentary rocks are formed from minerals deposited in layers, beneath the seas or on land. One group contains minerals from the breakdown of older rocks which can be subdivided on the basis of the size of their particles. The finest are clay-sized (i.e. less than 0.002 mm) accumulating as **mudstones** or **shales**. Slightly larger particles form siltstones. Sand-sized particles (commonly quartz grains) form **sand-stones** and gritstones. Larger rounded pebbles make up **conglomerates** and angular fragments form **breccias**.

Two processes change loose sediment into harder rock. The first is **compaction** where the weight of later overlying sediments squeezes the layers below. The second is

cementation when individual grains are glued together by iron oxides, calcareous or siliceous material. The variety of this cement contributes to the range of colours found in sedimentary rocks. An example is the transformation of rounded beach pebbles into a conglomerate.

Sedimentary rocks can also be formed by the accumulation of dead organic matter. Marine organisms can produce shells and skeletons of calcium carbonate. When they die these fall to the sea bed and accumulate as **limestone**. Examples in Britain include chalk, oolitic limestone and carboniferous limestone. Plants in swamps, deltas and estuaries may not rot away when they die. Such organic material after compaction can form lignite and coal.

As sediments accumulate, remains of plants and animals can be incorporated in their layers (**strata**). Unravelling the evolution of animals and plants from their fossil remains has produced and dated the *geological column*, table 4.1. (This shows the names of the eras and periods.)

table 4.1   The geological time scale

ERA	PERIOD	EPOCH	Age, in millions of years
CENOZOIC	QUATERNARY	Holocene	
		Pleistocene	2
	TERTIARY	Pliocene	7
		Miocene	26
		Oligocene	38
		Eocene	54
		Palaeocene	65
MESOZOIC	CRETACEOUS		135
	JURASSIC		200
	TRIASSIC		240
PALAEOZOIC	PERMIAN		280
	CARBONIFEROUS		370
	DEVONIAN		415
	SILURIAN		445
	ORDOVICIAN		515
	CAMBRIAN		590
PRE-CAMBRIAN			4600 Origin of earth

The third category is **metamorphic rock**. Temperatures rise 3°C for every 100 m deeper in the crust. Pressures also increase; at a depth of 24 km, for example, the pressure is 7200 times that on the surface! With temperature and pressure conditions like this it is not surprising that sedimentary and igneous rocks are altered, or *metamorphosed*, as they are buried and as the earth's crust moves. Minerals can become lined up in the same direction (foliated), fine-grained sediments like mudstones can be changed into slates, and limestone into marble, for example.

table 4.2   Rock classification

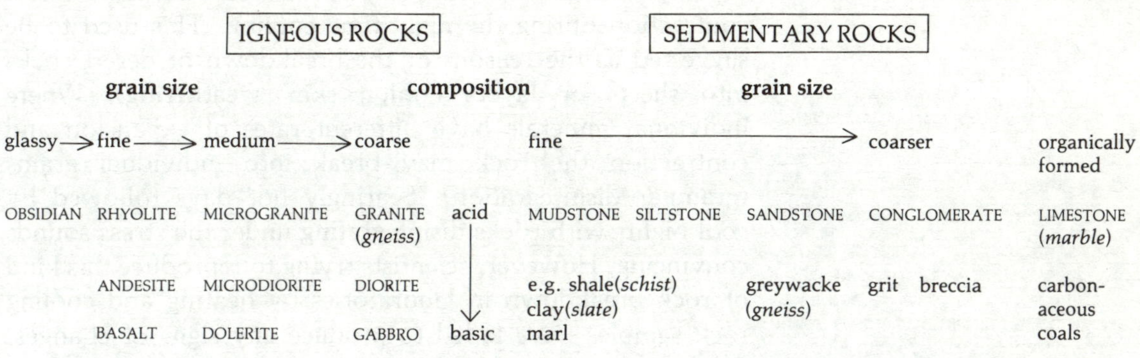

IGNEOUS ROCKS					SEDIMENTARY ROCKS			
grain size			**composition**		**grain size**			
glassy → fine →	medium →	coarse			fine ——————→	coarser		organically formed
OBSIDIAN RHYOLITE	MICROGRANITE	GRANITE (*gneiss*)	acid		MUDSTONE SILTSTONE	SANDSTONE	CONGLOMERATE	LIMESTONE (*marble*)
ANDESITE	MICRODIORITE	DIORITE	↓		e.g. shale(*schist*) clay(*slate*)	greywacke (*gneiss*)	grit breccia	carbonaceous
BASALT	DOLERITE	GABBRO	basic		marl			coals

SOME METAMORPHIC ROCKS altered sediments and igneous rocks are shown in italic

What are *the characteristics of rocks which influence their breakdown and movement?* The first is their chemistry i.e. what minerals they are made of. Secondly, there is the existence of pore spaces as well as cracks, fissures and crevices such as the bedding planes of sedimentary rocks. These allow air and water to penetrate. Thus both chemistry and cracks (i.e. structure) affect how 'hard' or 'soft' a rock is. A third factor is how rocks are arranged in relation to each other. However, before we look at rocks and scenery and at how rivers, glaciers and waves produce landforms, we need to consider how rocks decompose, decay and move.

What are the three factors which influence rock breakdown and movement?

Rock destruction and decay

At the surface rocks are subject to the effects of air and water which breaks them up and transforms them into new compounds. Rocks break down in various ways into blocks, sheets or individual grains.

figure 4.3   The signs of rock destruction in Utah. In this arid environment there are few plants to conceal how the rocks can break down as sheets.

Mechanical weathering

During the day the sun's rays pour onto the surface; this energy is absorbed by the rocks so their temperatures rise. At night the rocks, acting like natural storage radiators, lose energy back to the atmosphere and their temperatures fall. In deserts, with few clouds to blanket the radiation in and out, daily temperature variations can be quite large. Surface temperatures of basalt in the Sahara are as high as 40°C during the day and as low as 2°C at night. In other climates such **diurnal** (day–night) temperature ranges are less extreme.

Temperatures change the volume of rock. The daytime expansion of the surface layers of rock contrasts with contraction during the night-time cooling. This used to be suggested as the reason for the breakdown of desert rocks into sheets or layers ('onion skin' weathering). Where individual minerals have different rates of expansion and contraction the rock may break into individual grains (granular disintegration). Searingly hot days followed by cool nights with rocks disintegrating under the stress sounds convincing. However, scientists trying to reproduce this kind of rock breakdown in laboratories by heating and cooling rock samples have failed to produce any signs of changes! Something else must be at work.

When temperature changes are linked with the presence of water the rock disintegration process becomes more lively. Table 4.3 shows the number of rock falls from a chalk cliff in

table 4.3   The number of rock falls from a chalk cliff in Kent and the frequency of frosts

Months	Number of rock falls from cliff	Number of days with air frost
Jan	9	8.5
Feb	6.5	7.5
Mar	8	4.0
Apr	1.2	0.2
May	0	0
Jun	0	0
Jul	0	0
Aug	0	0
Sep	0	0
Oct	2	0
Nov	5	1.2
Dec	9	5.0
	average per month 1810–1970	*average per month 1956–65*

Kent. It also shows the number of days when frost has been recorded, a pattern which mirrors the first one.

Imagine a rock with fissures and cracks which liquid water can seep into. When the temperature falls below 0°C this water will freeze. If freezing is rapid a skim of ice will form across the entrance to the crack and seal it. Imagine the cooling continuing; deep within the crack water will finally turn to ice. When it does so its volume expands by 9 per cent and it can exert tremendous pressure on the rock, trying in the process to widen and deepen the crack. This process of rock disintegration is called *frost shattering*.

Ice crystals can also form in the tiny pore spaces within the rock. When this happens ice adds to its disintegrating power by pulling still unfrozen water to the crystals from surrounding rock pores, Such enhanced *crystal growth* appreciably weakens the rock.

One way to discover how easily various rocks break down is to take rock samples into a laboratory and expose them to **freeze-thaw** cycles (i.e. temperatures ranging above and below freezing). Figure 4.4 shows the results. Various rocks have different degrees of susceptibility to freeze-thaw attack. The granite is massive, with its crystals joined tightly together and with little chance of water penetrating. Chalk on the other hand has lots of pore spaces and tiny fissures, both ideal for ice crystal growth.

The results of freeze-thaw are usually sharp-edged, angular rock fragments. When a cliff is attacked, the loosened rock debris falls to its base where it can pile up as **scree** (fig. 4.5).

Freeze-thaw weathering

Examine fig. 4.4.
1 Describe how the various rocks differ in their freeze-thaw breakdown.
2 Explain these differences.

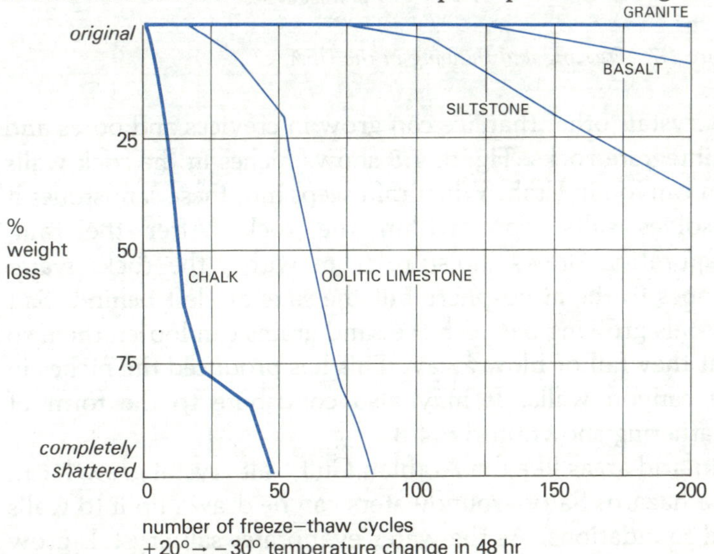

figure 4.4   *The susceptibility of five different kinds of sedimentary and igneous rocks to frost shattering*

figure 4.5   *Frost shattered cliffs and scree slopes below, Moraine Lake, Alberta, Canada*

49

figure 4.6 *Felsenmeer: a spread of angular frost shattered boulders in north Norway*

When the frost shattering occurs on flatter ground, **felsenmeer** (a sea of angular rocks) is produced (fig. 4.6).

You might think of the kind of environment where this freeze–thaw process occurs – areas where temperatures fluctuate around 0°C, and water alternately freezes and thaws. Low latitudes (the tropical lands) are non-starters as the temperatures rarely fall low enough. Arctic environments are cold enough, but it stays frozen for much of the year. Mid-latitudes are the most likely places since they have temperature fluctuations around freezing in autumn and spring. Figure 4.7 shows the distribution of freeze–thaw cycles in the USA.

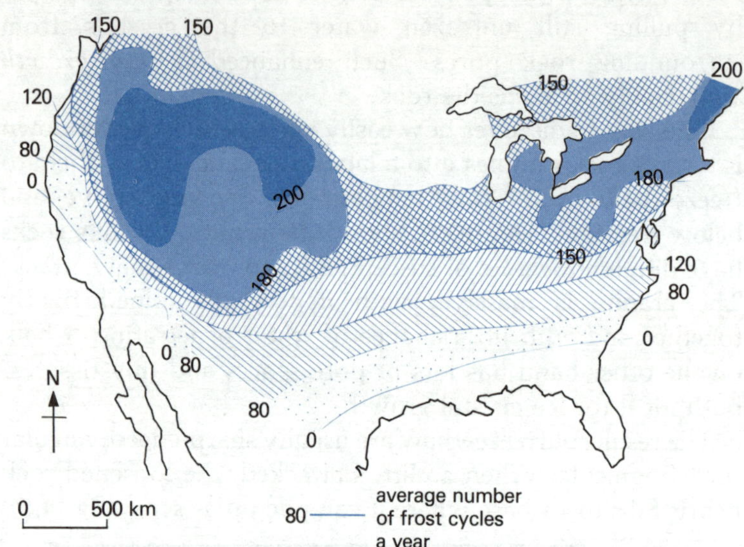

figure 4.7 *Freezing and thawing in the USA*

Crystals other than ice can grow in crevices and pores and disintegrate rocks. Figure 4.8 shows niches in the rock walls of a canyon in Utah. When rain seeps into these sandstones it dissolves salts from within the rock. After the rain, evaporation draws moisture from within the rock, water escapes to the atmosphere but the salts are left behind. Salt crystals growing between the sand grains can loosen them so that they fall or blow away. This has produced the niches in the canyon walls. It may also contribute to the form of weathering shown in fig. 4.3.

In arid areas like the Arabian Gulf, salt crystal growth can be a hazard. Salty groundwaters can be drawn up into walls and foundations. As the water evaporates salt crystals grow and can break up masonry and cement, leading to foundation failures.

50

figure 4.8   Weathering niches. The walls of the Canyon de Chelly in Utah are etched with hollows. The one at the centre, 50 m up the canyon wall, is occupied by the White House, an Anasazi Indian village abandoned more than 500 years ago.

Many rocks were originally formed beneath the surface. When the overlying rock is stripped away the removal of their weight allows the rock to expand or dilate. In rocks like granites, quartzites and thickly bedded sandstones, such *pressure release* leads to the growth of a series of fractures parallel to the ground surface. Huge slabs, sheets and **spalls** of rock are the signs of this kind of process (fig. 4.3).

All the processes described above leave the rock chemically unchanged. They are therefore examples of *mechanical weathering* processes. In nature these operate at the same time as chemical processes.

Chemical weathering

Visiting the country's cathedrals you might have found them shrouded in scaffolding and appealing for repair money. Statues and carvings decaying after a few hundred years' exposure to the atmosphere are signs of processes involving chemical weathering. Although discussed separately here, chemical processes act with mechanical processes to weather rocks.

Describe the ways in which rock can weather without its chemical composition being altered.

The carbonation reaction
$CaCO_3 + H_2CO_3 \longrightarrow Ca(HCO_3)_2$
i.e. limestone plus water and carbon dioxide (carbonic acid) produces the soluble calcium bicarbonate

The weathering of feldspar
$2KAlSi_3O_8 + 2H_2CO_3 + 9H_2O$
i.e. potassium feldspar plus carbonic acid and water
$\longrightarrow Al_2Si_2O_5(OH)_4 + 4H_4SiO_4 + 2K^{++} + 2HCO_3^-$
i.e. giving kaolinite (a clay mineral), silicic acid, potassium ions and bicarbonate ions

Rocks are composed of a variety of minerals. These are exposed to the atmosphere and to the effects of water and oxygen and other gases. One familiar weathering change is *oxidation*. Unprotected iron in dry air can remain clean and bright, like the remains of the North African tank battles of World War II. When moisture is present the iron rusts or oxidizes. There is always dissolved oxygen in rainwater and dew, which can change the ferrous iron in minerals to the more oxidised ferric state. Iron is a widespread element in the earth's crust and this oxidation–reduction process is the reason for the red or yellow coloration of much weathered rock debris and soil. The change destroys the original mineral structure and the rock is decayed and weakened.

Rain also picks up carbon dioxide as it falls through the air and, more importantly, as it trickles through the soil. Carbon dioxide turns the water into a weak acid, known as carbonic acid. This attacks limestones in the process of *carbonation*. Where limestone is very pure (i.e. almost entirely $CaCO_3$) there is virtually no residue; the water carries away the product of this kind of weathering.

A second, more complicated, example of how minerals and water react occurs in igneous rock weathering. Granite contains mica, quartz and feldspar. Let's look at what happens to just *one* of these – the feldspar. This is more complicated than the carbonation reaction affecting limestone but it is just one of the reactions affecting granite. Its other minerals are also attacked. Not all minerals are equally susceptible, however. The quartz, for example, is relatively resistant but when its neighbours decay, soften, expand or wash away it is very much a case of the chain (the granite) being as strong as its weakest link. The end products of granite weathering are therefore clay minerals, like kaolin or china clay which is a valuable economic resource, unaltered or residual particles (like the quartz grains) and material washed away in solution.

Icelandic scientists have measured the amount of dissolved products in streams draining away from basaltic lavas. They found three times as much material being removed from the older flows covered with plants. This is just an example of the influence of organisms on chemical weathering. In humid environments like Britain bare rock is only rarely exposed at the surface – it is usually mantled with a mat of soil and vegetation. Sometimes roots may physically break up the rock (fig. 4.9) but far more important is the chemical attack on rock which plants contribute to. As they live they draw up dissolved nutrients, as they die these return to the soil. A

figure 4.9   Root power! Massive Carboniferous limestone in Burrington Combe, Somerset, split by tree roots exploiting the lines of weakness in the rock's joints and bedding planes.

whole complex of chemical reactions occurs, 'clawing up' some minerals and decaying others.

1 Describe the ways in which rock decays chemically.

2 Make a list of the various ways in which rock can disintegrate or decay and describe what the products are like.

How material moves on slopes

In the Vaiont disaster rock movement was a *sliding* one and very *quick*. The clay bands and the tilting of the rock layers favoured movement, but the *trigger* for the disaster was heavy rain. Other rapid **mass movements** may be triggered by earthquakes. One of the most disastrous took place in Peru in 1970 on the 6700 m high Huascaran, a peak in the Andes. A huge volume of rock (100 000 000 m³) *fell* from the north face of the mountain. Hitting the ground, the rock pulverised and rushed with a noise like thunder fifteen kilometres down a gorge, travelling at 400 kph. Just three minutes after the rockfall began the town of Yungay was obliterated and 40 000 people were killed. Rapid mass movements are at their most dramatic in high mountain areas with steep slopes.

figure 4.10   *Evidence of movement. The plastic tube has been bent by five years of solifluction in Labrador–Ungava, Canada. The horizontal marks are 30 cm apart.*

Not all mass movements of rock debris are as rapid and not all movements involve sliding or falling. Five years before fig. 4.10 was taken, a plastic tube had been driven vertically into the clay on a valley side. When the site was re-surveyed the top of the tube was discovered to have moved 25 cm downslope. A pit was dug alongside the tube to see what shape it was. The bending shows us that the surface layers have moved more rapidly than the lower layers. In other words, a *flow* had taken place. The clay had behaved like thick oil or syrup on a sloping plate – its lower part slowed by friction with the base and its upper part moving more freely downslope.

The photograph (fig. 4.10) was taken in northern Canada where in winter the ground is frozen. In summer, after the snow has melted, the ground is saturated with water. The clay begins to surge and flow downslope. This kind of soil flowing process is known as **solifluction**. During the summer, as the clay dries out the movement slows. Solifluction also occurs in other areas. Lowland Britain contains many 'fossil' solifluction deposits. During the recent glacial period many valley sides and floors became mantled and plugged with solifluction deposits. These deposits are known as combe or head (fig. 4.11).

When you first looked at fig. 4.12 you might have wondered what it is. In fact it is a vertical air photograph of a snow covered suburb of Anchorage, Alaska, with snow-cleared roads and diagonal shadows cast by pine trees. You

figure 4.11   VI Form students examining a 'fossil' solifluction deposit (head) on a hillside in Devon

figure 4.12   Turnagain Heights, Anchorage, Alaska. A vertical air photo (for explanation see text).

probably found it difficult to interpret the chaotic area at the top of the photograph. It was taken after the Good Friday Alaskan Earthquake of 1964. This massive earthquake jarred half of the State and cracked ice on rivers and lakes over 250 000 km², four times the area of Ireland! Anchorage was 100 km from the focus of the earthquake but on the photograph we can see that the 22 m high bluff on which the houses were built has collapsed as a crumbled, dishevelled mass into the ice-speckled arm of the sea. Below is a transcript of R. Pate, the announcer on duty at KHAR Anchorage, who described his feelings during the quake.

A transcipt from station KHAR Anchorage Alaska during the Good Friday Earthquake. These are the words of the announcer in the studio as the earthquake happened:

'Maa-uhn! – I'm not faking a bit of this – I'm telling you the whole place just moved like somebody had taken it by the nape of the neck and was shaking it. Everything's moving around here! . . . Oh I'm shaking like a leaf . . . I wonder what it did to the tower . . . Boy! the place is still moving! You couldn't even stand up when that thing was going on like that – I was falling all over the place here. . .I'm telling you, this house was shaking like a leaf! The picture frames – all the doors were opened – the dishes were falling out of the cabinets – and it's still swaying back and forth – I've got to go through and make a check to make sure that none of the water lines are ruptured or anything. Man, I hope I don't live through one of those things again . . .'

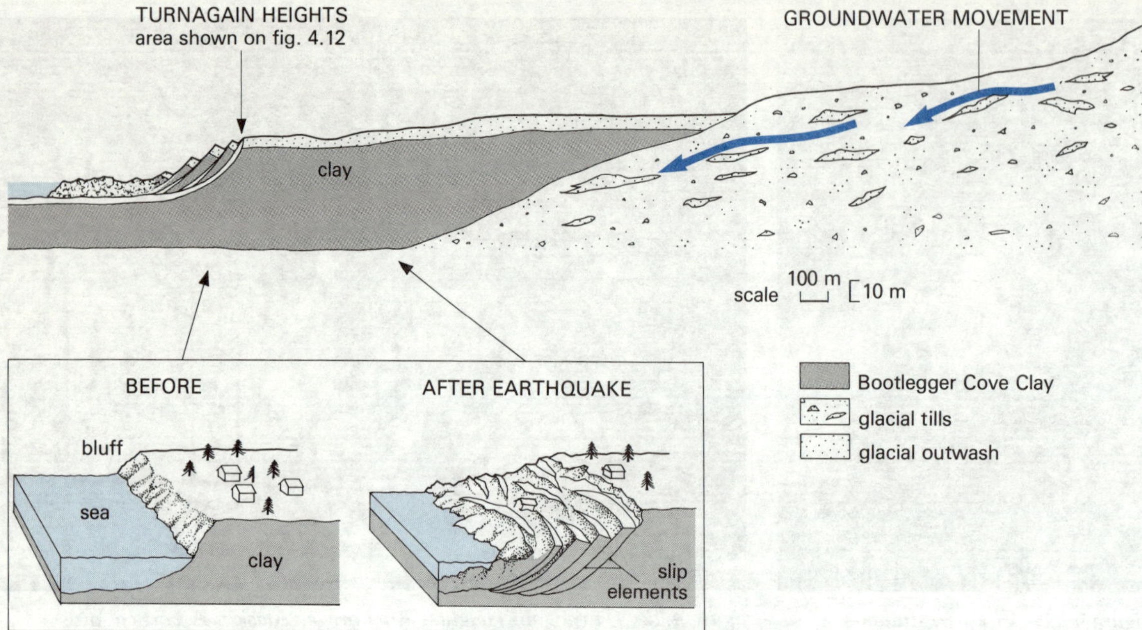

figure 4.13   A section under the Turnagain Heights area, Anchorage, Alaska, showing the effects of the earthquake

This material has moved in a slightly different way from the slides and flows mentioned earlier. The left-hand end of fig. 4.13 shows curving 'slip planes'. This type of mass movement is known as *rotational slip* – if you look carefully at the photograph you can see the slips overlapping. Their intact tops are still snow mantled and in places covered with parts of houses.

The Vaiont rockslide was analysed in terms of 'setting' and 'trigger'. The Alaskan 'trigger' was the severe shaking administered by the earthquake – but how did it work here? If you look at fig. 4.13 you can see that the houses were underlain by a deposit, known locally as the Bootlegger Cove Clay. This clay had been deposited in sea water and its tiny mineral particles were held together by salt, acting as a kind of 'electrolytic glue'. At the end of the Ice Age the land rose above sea level and, as fig. 4.13 shows, water from inland began passing through it and started flushing out the salt.

This created a potential hazard as the clay had become *thixotropic*. Non-drip paints are like this; solid in the tin, but as soon as the brush goes in its fibres stir up the paint particles which become liquid. The Bootlegger Cove Clay had this peculiar quality, it was a 'quickclay'. As it was agitated by the shaking of the Good Friday earthquake, it lost its cohesion and shear strength. The bluff bordering the sea 'failed' with concentric slices collapsing behind it.

'Quickclays' are common where uplift after the Ice Age has exposed clays to flushing. Areas around the St. Lawrence and in Scandinavia have suffered from this dramatic form of sudden slope failure. Rotational slip can often be seen in Britain, especially at the coast where the sea removes the basal support of the cliff. The slope fails on a curving plane leaving amphitheatre-shaped scars on the hillside. On a small scale you may have seen this on motorway embankments.

So far we have considered sliding, flowing and rotational slipping as various ways material moves downslope under the influence of gravity. It can however move less dramatically as **soil creep**. All soil is ultimately rock on its way to the sea and as Professor Kirkby said 'the creep mechanism is its large and slow magic carpet'. It may not be very dramatic, but think of how many times a year a British hillslope has rain falling on it and then it dries out, how many times soil freezes and then thaws. Both these cause the carpet of soil to heave upwards (i.e. at right angles to the slope). When it dries (or thaws) it gets pulled back vertically by gravity. Plants sway in the wind, their roots tug and disturb the soil particles which settle back vertically. Lowly worms and rabbits loosen soil, even geography students on field courses trample it! All these disturbances result in a tiny net movement downslope known as soil creep. There may also be flowing movements in waterlogged clays or even tiny scale rotational slips. They are undramatic but very important when you think they are continuous and occur widely. There is no need to wait for earthquakes to move material downslope!

In studying how rock material moves downslope we must not forget the effects of rainsplash. Raindrops falling onto a bare soil surface act like tiny bombs, dislodging particles with their impact. The dislodged particles fall vertically under the influence of gravity. This means that there is an overall movement of soil downslope.

Invisible movement: solution

Usually it is only when natural processes achieve something spectacular that they make the headlines, as you can see from the extract overleaf.

The explanation in the last paragraph of the extract is only part of the story however. Much of Florida is underlain by limestone and it was the **solution** and removal of this rock which created the caverns underground whose roofs collapsed in such a spectacular way.

figure 4.14   *Signs of soil movements in the Black Mountains, Wales. The slope is lined with terracettes.*

Make a table, one side headed 'background setting', the other 'trigger factors'. Describe (using diagrams if you wish) the factors and setting for the mass movements described so far in the chapter.

Town's sinking feeling

From Harold Jackson in Washington

THEY recently built a new swimming pool in the small town of Winter Park, Florida. It is three feet deep at one end and 175 feet at the other. That was not the original design: half the pool has been gobbled by an enormous and expanding hole.

The hole has also consumed two businesses, a house, five Porsches, several trees, and a large lump of suburban road. It was still expanding yesterday.

It originally manifested itself on Friday to Mrs Mae Rose Owens. As she looked out of her window she saw a nearby tree getting shorter by the minute.

"All of a sudden, the earth just opened up and, ploop, down this tree went. I couldn't believe it."

By 4 am on Saturday, Mrs Owens had seen her home vanish, followed shortly afterwards by the premises of a German car service company and a printing firm.

By Sunday, the hole had consumed two of the four lanes of Denning Avenue, and was homing in on three more houses. Yesterday it measured 1,200 feet across and 175 feet deep and, according to Mr Jim Smoot, of the US Hydrological Survey, will probably go on growing for several weeks. The local fire department reported about 50 feet of water in the bottom.

The hole seems to be one of the less expected consequences of the continuing drought in much of the United States. According to local experts, underground streams in the region have dried out, contracting the soil.

An extract from the Guardian, May 1981. During 1981 rainfall was low in Florida and the watertable in the limestones lowered so that cavern collapse could occur.

The term limestone covers an enormous range of rocks, of different colours, ages, strengths and degrees of fissuring. A simple definition is that limestone is any sedimentary rock which contains more than 50 per cent calcium carbonate. The basis of limestone solution is carbonation, when calcium in the limestone is dissolved in water (see page 52). This can be detected by measuring the hardness of water (expressed as so many parts of calcium per million parts of water, abbreviated as ppm). Those of you living in hard-water areas will have seen signs of it in the scaling of kettles.

Pure water can dissolve enough limestone to produce a hardness of 9 ppm. Sampling springs and streams in limestone areas however gives hardnesses of up to 250–300 ppm. We have a problem – more powerful solution is occurring somewhere.

Part of the answer is that rainwater is not pure water. It has dissolved carbon dioxide from the air, forming dilute carbonic acid. Rainwater is therefore *more aggressive* than pure water. In Britain it is aggressive enough to carry away about 75 ppm of calcium. This helps but the problem remains – how has that extra calcium been dissolved?

Carboniferous Limestone is a massive rock with well spaced jointing and bedding lines (fig. 4.9). These have been enlarged by the solution process so that interconnected *cave* passages have been produced, it is therefore possible to sample water within the rock (fig. 4.15). The diagram shows

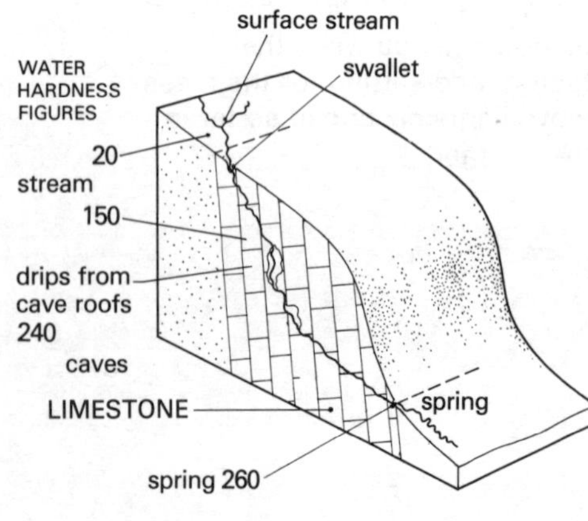

figure 4.15 Water hardnesses in a limestone area

figure 4.16   A limestone pavement at Malham, Yorkshire

figure 4.17   A vegetated limestone pavement

that water dripping from cave roofs (after passing from the surface through the rock towards the caves) has very high hardness figures. Such water must have become much more 'aggressive', or *acidic*, than rainwater. Somewhere along its route it must have picked enough carbon dioxide to make it a more powerful acid able to dissolve more limestone.

An answer can be found by looking at fig. 4.16 which shows a **limestone pavement**. Its surface is fretted by a network of crevices. Figure 4.17 shows the same kind of plateau edge situation with virtually horizontal strata, but here the rock is still covered by soil and vegetation. If the trees were felled by man, the grass overgrazed by his sheep and cattle and the soil then eroded away, the underlying surface of the limestone would look like the pavement in fig. 4.16. So does the answer to our problem of solution lie in the soil?

Soil air is different from the atmosphere, containing more carbon dioxide – as much as 10 to 10 000 times more. Rainwater passing through absorbs this carbon dioxide and becomes more acidic. It is in the soil, then, that water picks up enough carbon dioxide to produce the kind of hardness figures detected in the caves and springs. The sub-soil zone is now believed to account for two-thirds of weathering. The water flowing through the cave systems actually dissolves only a tiny amount.

The origins of dry valleys

Limestone scenery

Figure 4.18 is an air photo of part of the Mendip Hills in Somerset and fig. 4.19 a sketch map of the area. The 300 m high area at the bottom of the photo is underlain by impervious Old Red Sandstone, so rain falling on it forms streams draining northwards. These disappear when they flow onto the carboniferous limestone. The sites of such disappearing streams are known as **swallow holes**. The streams continue their journey underground through a network of crevices, pipes, larger passages and caves before reappearing at the edge of the limestone. The network of underground drainage can be complicated. Water-tracing experiments have used dyes and spores, inserted at the swallow holes, to discover connections (fig. 4.19). The underground water is using a network created by solutional attack (on the kind of bedding and jointing cracks shown in fig. 4.9, actually taken in Burrington Combe) and normal running-water erosion.

The air photo also shows a line of small *surface depressions*. These may have been formed by solution from the surface being concentrated at weaker points in the limestone, where joints cluster for example. They might also form by the *collapse* of a cave passage, a less dramatic example than the one in the newspaper report on page 58.

The largest features are the **dry valleys**; their shapes suggest they were cut by running water but they do not contain streams today. There are a number of theories about their origin.

1. Underground drainage takes time to develop. Until then surface runoff would have taken place, carving the usual network of river valleys.

2. The water table in the limestone area may have been higher in the past, either because the area was wetter or because the land surrounding the limestone upland was higher.

3. However, in the case of Burrington Combe its most recent use occurred in the Ice Age. At that time, as recently as 20 000 years ago, this area of Somerset had a tundra-like climate, as we know from the traces of Stone Age hunters who sheltered in the caves. The ground would have been permanently frozen (permafrost). This would have made the limestone impermeable so that surface streams would have flowed, very likely quite powerfully too as the winter's snow melted in early summer. As the climate warmed, the permafrost would have melted and the spring snow melt floods decreased.

figure 4.18   An air photo of part of the western Mendips in Somerset

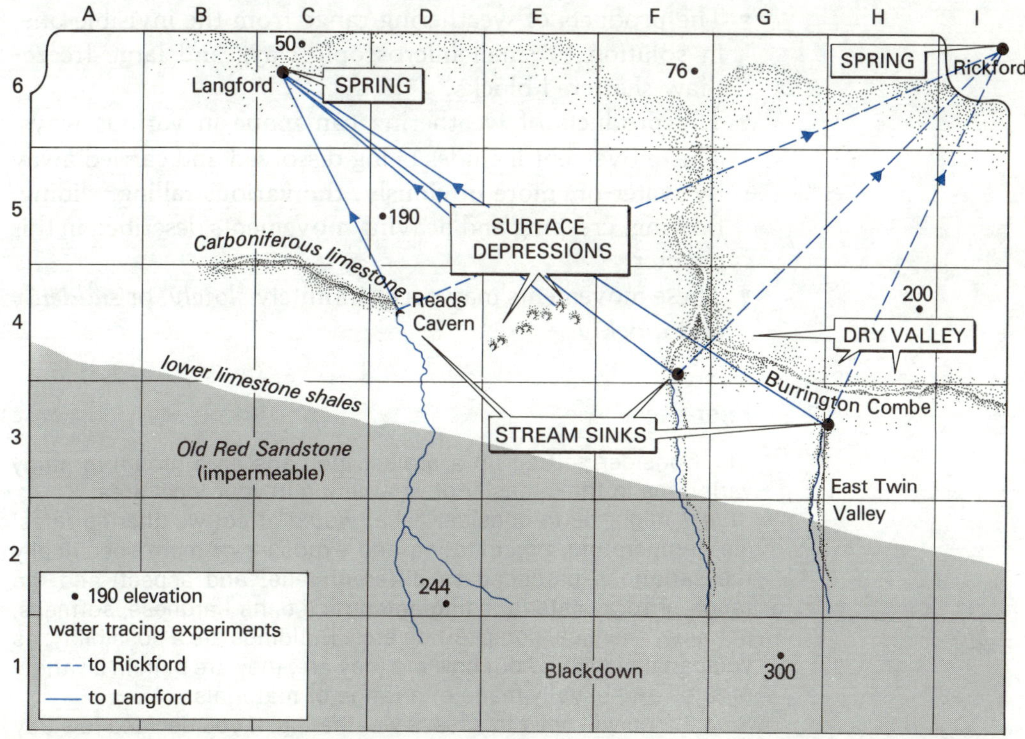

figure 4.19   Some features of limestone scenery in the Burrington area of the western Mendips, Somerset. The area is that shown in the air photo above.

Scarps are discussed in chapter 8

Limestone areas, therefore, have a range of features reflecting the solution processes and underground drainage. Its not surprising that they have a special name – **karst landscapes**, named after an area of Yugoslavia where the limestone scenery is particularly impressive.

Chalk is chemically similar to the kinds of limestone mentioned above and it is as liable to solution. However the detail of chalk landscape is very different, its hillslopes are more rounded and cave systems are rare. The chalk, unlike massive limestones, has many tiny fissures for water to move through. This means that solution is not concentrated along a few major cracks which can widen into caves. It is also mechanically weaker and could not hold a cave passage open. (This reminds us that when we consider a weathering/erosion process, like solution, we do need to remember the kind of rock it is affecting.)

Summary

- Some of the ways in which rocks disintegrate and decay have been introduced in this chapter.
- The two weathering processes (mechanical and chemical) interact.
- The products of weathering range from the invisible ones in solution to small microscopic clays and large freeze-thaw shattered blocks.
- The products of weathering can *move* in various ways. This movement includes being dissolved and carried away by water or, more obviously, the various falling, sliding, flowing, creeping and heaving movements described in this chapter.
- These movements may occur infinitely *slowly*, or *suddenly* and *quickly*.

Exercises

1 Consider setting up a project (perhaps as a group) to study variations in the intensity of weathering in your local area.
Step 1 might be to consider 'what would affect weathering rates' (i.e. temperature, precipitation and exposure or more specifically the variations produced by different relief and aspect) and the range of rock material being acted on (i.e. its hardness, softness, porosity, chemical composition etc.). Buildings are a possibility, as you can usually find out how old they are, they are built in a variety of sites and usually made of a range of materials.
Step 2. You will need to assess weathering. If you decided to study inscriptions, a six-point scale could be: 1 unweathered, 2 slight weathering (edges and letters slightly rounded), 3 moderately

weathered (rough surfaces), 4 badly weathered (inscriptions difficult to interpret), 5 bad weathering (any letters virtually indistinguishable), 6 extreme weathering.

Step 3. You need to decide where to go to survey. This step is an important part of your research design, so it is worth buzzing ideas around before you agree on the detail.

Step 4. Process your results and produce your findings. This will involve thinking about tables and graphs to show how weathering varies with factors like relief, exposure or material.

2   If you were given the task of advising on the route of a new motorway, what kinds of information would you need before you could warn of geological hazards?

3   Describe how limestone solution is measured. What are the landforms produced?

4   Look at fig. 4.20 *and* its caption. The Tribunal of Enquiry after this disaster concluded that the Aberfan disaster could and should have been avoided. List the reasons which would back up this statement.

figure 4.20   The Aberfan flow slide. On October 21st 1966 a coal tip collapsed killing 147 people, including 116 in the local school. The scene shows Aberfan in the Taff valley and the colliery tips above it. 107 000 m³ of material slid and flowed at up to 32 kph down into the valley. The enquiry tribunal found that water percolating through the sandstone under the tip issued as a spring. This had removed material from the toe, or bottom, of the slope. Heavy rainfall then saturated the loose material of the tip. The flow slide moved more quickly than previous smaller movements. This disaster focused attention on the other 1300 tips in the South Wales coalfield.

Rivers and landscape change

The cycling of water was introduced earlier in chapters 2 and 3, whilst rock decay, destruction and mass movement were the focus of chapter 4. It is now time to weave these together and discover the landforms which running water produces. We begin with the middle course of the river. A curious starting point, but one which shows the interplay of different processes at work within and around the river.

Floodplains and river work

From time to time rivers flood. Their level rises, their banks are overtopped and water spreads across any surrounding lower ground as you can see in fig. 5.1. The area affected by such periodic flooding is known as the **floodplain**.

What happens on floodplains?

In physical geography it is difficult to find out how processes work today, what has happened in the past and what might happen in the future. These questions are particularly important for floodplains. Their flat land, floored with clays, sands and gravels, has been attractive for a range of purposes. It is quick and easy to build homes and factories across them, run roads and railways along them, grow crops and keep livestock on them and lay out sports grounds on them. It has also been easy to dig into them for sands and gravels, as has occurred in the Thames valley west of London.

Deciding how to use floodplain land is therefore a fairly central *problem* for land-use planning. Part of the problem is that these land uses *conflict*. The other part of the problem is how to map *floodplain limits*. A simple definition of floodplains is that land close to the rivers which gets flooded. This is vague; some parts get flooded every year, other parts every ten years and higher areas only every hundred years or so. Such a hydrological definition of a floodplain, therefore, needs mention of the likely *frequency of flooding*. Flood risk may help decide land use. Yearly flooding of farmland may

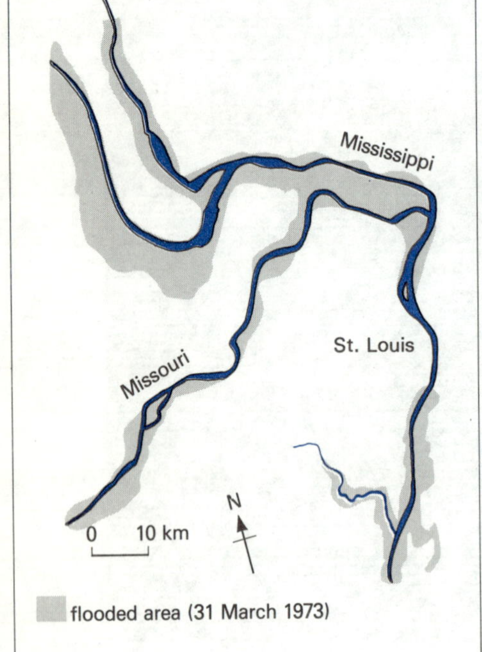

figure 5.1   *A map of the March 1973 Mississippi flood. Sometimes river flooding is on such a large scale that the quickest way to discover the extent of the hazard is from satellites. The narrowing of the flooded area around St. Louis shows the effectiveness of the city's flood protection embankments.*

not cause much damage, it might even be a benefit! The risk
of flooding warehouses every ten years might be acceptable
but would not be acceptable for homes, when a one in fifty
year chance is a more 'accepted' risk.

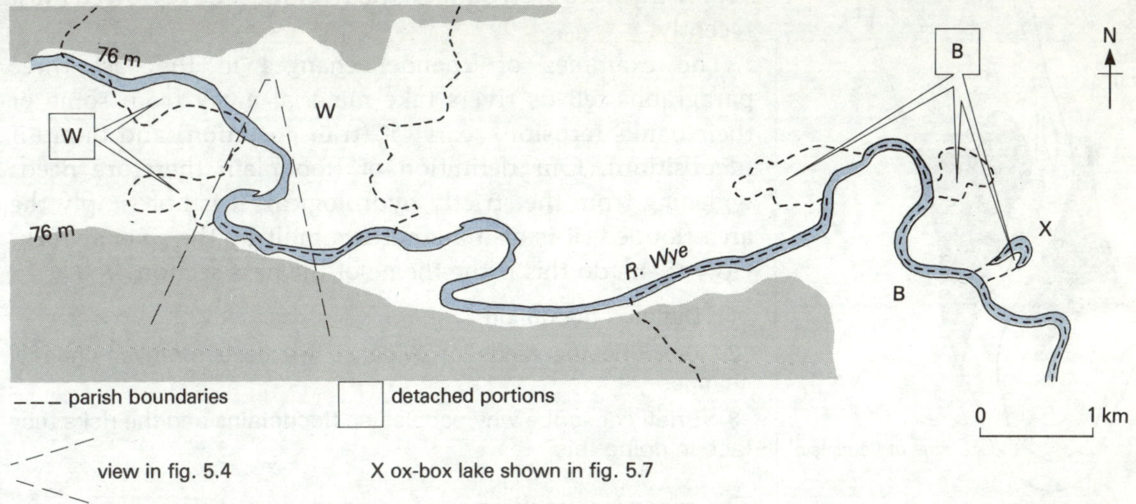

figure 5.2   Part of the River Wye's floodplain. (This section of the river is 25 km downstream from Erwood whose position is shown in fig. 3.7).

Answering the question 'what happens on floodplains'
involves us in historical research. As rivers form natural
boundaries, it is not surprising that many medieval parish
boundaries followed river lines. Figure 5.2, however, shows
that since the boundaries were drawn, the River Wye has
changed its course, leaving parts of the parishes on opposite
banks of the river.

The Mississippi–Missouri mentioned already (fig. 3.21) is a
much larger river system than our other examples, draining
1 500 000 km² in the centre of North America. The Missouri is
known as 'Old Misery' on account of the trouble it caused
farmers and travellers over the years. George Fitch, writing at
the end of the steamboat era, wrily noted that

> 'a steamer that cannot on occasion climb a steep clay bank,
> go across a cornfield and corner a river that is trying to get
> away, has little excuse for trying to navigate the Missouri!'

The Little Missouri River is a tributary of the main river. In
North Dakota it is a shallow muddy river. Its banks and
valley floor are covered with cottonwood trees. B. Everett, a
geomorphologist, wondered if these might be a clue to the
river's history. Each year as the trees grew they added a new
growth ring to their trunks and stems. By boring holes and
counting the number of growth rings the ages of the trees
could be discovered. Everett surveyed the Little Missouri's

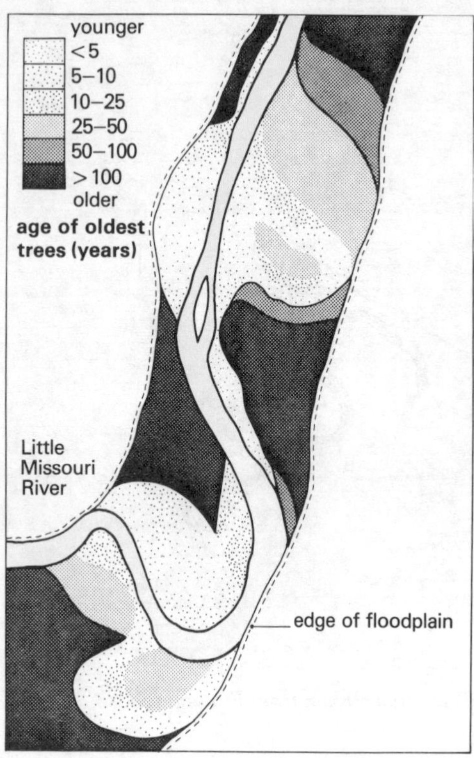

figure 5.3   The age of trees on the floodplain of the Little Missouri River. (The Mississippi-Missouri is shown in fig. 3.21)

valley floor and mapped how old the trees were. Figure 5.3 shows that in some places the river was bordered by old trees, in other parts by young ones. It has abandoned parts of the floodplain, which have then become colonised by trees. Where the trees are youngest the river must have flowed most recently.

The examples of channel changes in the last three paragraphs tell us rivers take material away from some of their banks (**erosion**), carry it (**transportation**) and dump it (**deposition**). Our definition of floodplain therefore needs widening from the strictly hydrological. It is not simply the area flooded, it is also a landform built by the river's work. How rivers do this is the theme of the next section.

1   Define a floodplain.

2   Describe the kinds of evidence we have for river channel changes.

3   Briefly describe why people use floodplains and the risks they face in doing this.

The meander machine

Figure 5.4 is a view of the River Wye. The river channel sweeps in a series of curves across the floodplain. These curves are known as **meanders**. Meanders are common features of floodplains so it will be useful to look at this one in more detail to see what is happening.

The relative speed of the water flowing around this bend is shown in part A of fig. 5.5. Notice how the faster flowing

figure 5.4   A meander on the River Wye. The location of this photo is shown in fig. 5.2.

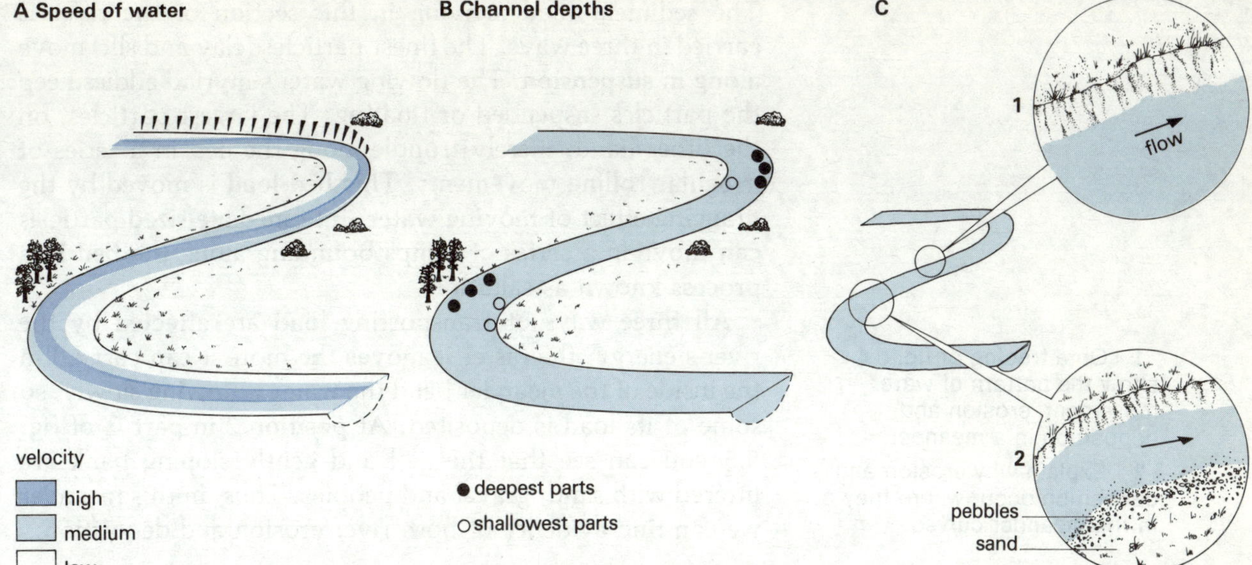

figure 5.5   *The dynamics of the Wye meander*

water hugs one bank and then swings across the channel to hug the outer side of the next bend downstream. The flow on the inside of the bends is much slower.

Part B of fig. 5.5 shows the water depth. Notice how the deeper parts, the pools, occur on the outside of the bends. This means the channel's cross section is asymmetrical at these points.

Part C of the diagram gives some detail of the edges of the channel. At position 1, the bank is steep and made of clay and silt with small pebbles set in it. This material is not that strong, so the near-vertical 'miniature cliff' suggests that it is a 'fresh' feature. There are also cracks, showing that the bank is tending to fall into the river. The outside of the bend is being *undercut*, the river is removing bank material. The river has a number of ways of eroding its bank and bed. At this part of a meander the water is moving relatively fast, and its impact on its channel side and bed can smooth them. This is known as **hydraulic action**.

This part of the Wye, in the middle of the basin, is receiving water from all the area upstream. Before it arrives in this section the river may have picked up clay, silt and sand and may be rolling larger gravel and pebbles along its bed. Such debris is known as **load**. This can hit against the bottom and sides of the channel and wear it away, the **corrasion** process. It is the combination of hydraulic smoothing and corrasion which has undercut and steepened the bank at position 1 in fig. 5.5.

Erosion

(Erosion processes are explained later)

Transport

67

The sediment load arriving in this section of the river is carried in three ways. The finest particles (clay and silt) move along in **suspension**. The flowing water's myriad eddies keep the particles suspended or floating. The largest particles, on the other hand, merely trundle along the bed in a series of irregular rolling movements. This **bed load** is moved by the dragging effect of moving water. Intermediate sized particles can move in a series of jumps bounding along the bed in a process known as **saltation**.

All three ways of transporting load are affected by the river's energy; the faster it moves the more it can carry. On the inside of the meander bend the water is flowing slowly, so some of its load is deposited. At position 2 in part C of fig. 5.5 you can see that the bed and gently sloping bank are littered with sand, gravel and pebbles. Thus, in this meander we can find evidence of both river erosion and deposition.

Working overtime

Looking at a river's quiet, meandering reach when it is not in flood is misleading. It is a different story in flood. The river's discharge increases, it deepens filling more of its channel and its flow is quicker. With these changes its ability to do work increases dramatically and the extra energy is used for erosion and transport.

During a flood, erosion on the outside bend of a meander can proceed more quickly. The river is flowing rapidly and more of its bank is exposed to erosion because the river is deeper. The faster flowing water is also carrying more load to wear away the bank.

When the flood is over, discharge falls, the river becomes shallower, it flows more slowly and the inside bends of the meanders become the site of fresh deposition. In these slower flowing waters (as can be seen in fig. 5.5) pebbles, sands and gravels are deposited to build a **point bar**.

When we consider a series of floods, increased erosion at the peak and deposition on the wane produce downstream *meader migration*. Figure 5.6 shows the gradual migration downstream of the meander as a result of erosion on the outer banks and point bar deposition on the curve's inner banks. The point bar deposits gradually fill the space occupied by the channel as it moves its way down the valley.

Besides eroding their banks rivers also erode their *beds*. As the meandering river swings across the floodplain, all parts of the rock underlying it are eventually liable to erosion. Meanders also widen valleys when they coincide with the

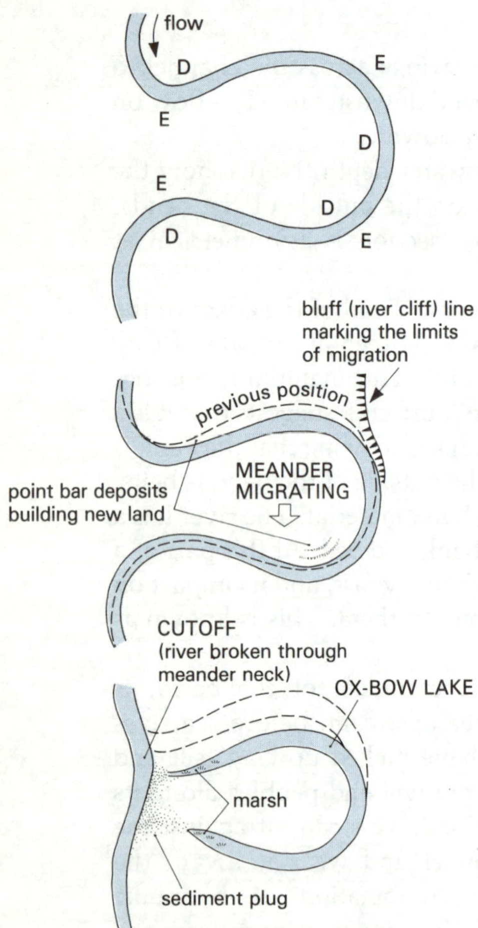

figure 5.6 *Meander migration*

slopes bordering the floodplain, producing a low 'cliff' line or **bluff**. The movement of meanders is therefore a part of the slow process of river valley widening and lowering.

Floodplains can be imagined as stores, places where sediment may be dumped temporarily on its way from the hillslopes to the sea. Most of the transfer in and out of the store occurs in floods. The idea of a floodplain as a store is illustrated in fig. 5.3. Here some material had been dumped recently and some had been stored for over a hundred years.

The meander migration, illustrated in fig. 5.6, may not be able to continue unhindered, as bank material varies: a patch of more resistant bank material may slow the process. The meanders upstream may then become more extreme in their curves as they slowly 'pile up'. This means that the meander may become so looped that only a narrow neck of land separates two meander cliffs. Such a neck can be pierced when the river is in flood, leaving the original course of the river as a cut-off loop. Its entrances become blocked with sediment, leaving a horseshoe shaped lake to mark the river's original course. Such a body of water is known as an **ox-bow lake**. This process is shown in fig. 5.7 at Letton (location in fig. 5.2) where the ox-bow lake is slowly filling with vegetation.

Describe the sequence shown in fig. 5.6 and explain how erosion, transport and deposition combine to produce meander migration.

figure 5.7   An ox-bow lake on the River Wye. The location is shown on fig. 5.2.

Using diagrams explain how an ox-bow lake is formed.

From the top or from the side? How deposits arrive in the floodplain store

When rivers overflow their banks they spread a relatively thin layer of water across the floodplain. Figure 5.8 shows what happens to the sediment load when this happens. Slow moving water 'rains down' sediment on the plain. The coarser and heavier load falls close to the river channel; in the quieter waters away from the fast-flowing channel smaller amounts of finer silts and clays are eventually deposited. You can see this 'sorting' if you mix soil and water in a jar and shake it.

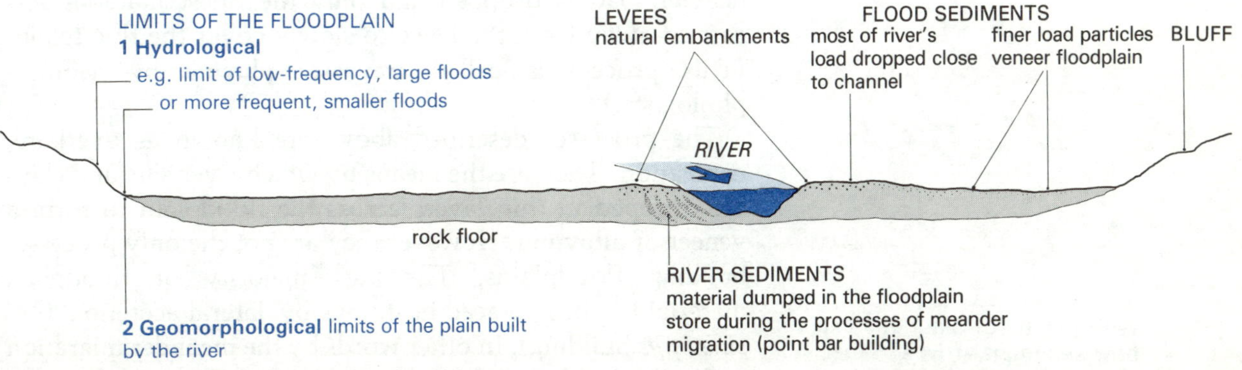

figure 5.8   Overbank deposits on the floodplain

figure 5.9   Crevasse splaying. This is a process occurring when river floodwaters breach a low or weak point in the levee embankments. Sediment charged floodwaters spew through the breach, leaving the coarser sediments close to the river and splaying out the finer material further away from the breach. This process is also a vital part of delta growth.

The first particles deposited once you stop shaking (i.e. like sediment-laden floodwaters once they leave the main river channel) are the larger ones. Depending on the texture of your soil it could take days for the finest clays to settle. Such a pattern of sedimentation gradually builds low embankments bordering the main channel. These natural embankments are known as **levees**.

In some places the levees will be lower, or weaker, and the sediment-laden waters of the larger floods can breach them. This situation is shown in fig. 5.9. Close to the breach the heavier load is dropped and only the finest material gets carried away from the levee to slowly cover the floodplain. (This process is called **crevasse splaying** by geomorphologists.)

The processes described above are known as overbank deposition. They are the means by which river silts and clays are dropped in thin layers across the floodplain to form a veneer of **alluvium**. However they are not the only processes building floodplains. The vast majority of floodplain material has been placed in storage by 'lateral accretion' (i.e. *sideways* building). In other words by the meander migration and point bar building mentioned earlier.

With the aid of diagrams explain how sediment arrives in the floodplain store.

70

Meanders and floodplains: summary

Meanders can be found in all sizes of streams. The larger the river, the larger the meanders. The proportions of their curves, however, remain similar. Why meanders form is a complex question. Experiments in laboratories, with straight channels moulded in sand and then filled with flowing water, have shown that the first changes are in the bed. Deeper **pools** are eroded and shallower **riffles** deposited. The pools then migrate sideways as the sinuous loop of the meander forms. Water in channels flows in a corkscrew manner and the eventual curving form of the meander seems to be a result of this.

If you look at maps or air photographs, you may have noticed that not all streams meander; some may have meandering reaches followed by straighter or irregular sections. If its banks are soft, every change in the amount of water flowing in the river alters the channel shape. Bank materials like sand, which lacks cohesion, or stickability, are especially prone to such continual reworking with every change in river discharge. On the other hand, if the bank and bed are in hard rock the river does not find it easy to shape its channel and meanders are rare. Meanders are therefore best developed when the banks are soft enough for floodwaters to mould them but strong enough to last until the next flood.

Braiding is another channel pattern produced by erosion and deposition (fig. 5.10), where a stream splits into different channels separated by shoals and islands. It is a common channel form where floodplain material is sandy (as it is in glacier meltwater streams) and where the valley gradients are too steep for meandering.

As we have seen in this section, the floodplain is the scene of various kinds of river work. Landforms are produced as the river erodes, transports and deposits material. We have also stressed that it is in times of flood that most of this work is done.

River erosion

Figure 5.11 is a view taken 10 km from the Wye's floodplain (shown in fig. 5.2). It shows a series of gulleys a few metres deep etched by running water into the flanks of the Black Mountains. This area has a precipitation of 1400 mm a year and the slopes are mantled with grass. In contrast, fig. 5.12 is from a much drier area in Utah where the gulley development is more dramatic. This kind of dissected terrain with many small valleys divided by sharp-edged ridges is often termed

figure 5.10   *A braided stream. The channel divides and joins.*

figure 5.11   *Gulleys on the flanks of the Black Mountains*

figure 5.12   *Gulleys dissecting horizontal sedimentary rocks in semi-arid Utah*

badlands. Both photos show the erosional downcutting effects of water flowing in stream channels.

The small Slapton stream in Devon was mentioned in chapter 3 when we first looked at the flood hydrograph. Attempts have been made to discover how much and what kind of erosion was happening in this basin. One way to do this was to measure the output of erosion products from the basin. The solid material was either leaving as bed load or as suspended load. Bed load was measured by sinking a box across the bed of the stream to trap any rock material moving along its bottom. This could be emptied and weighed regularly. Measuring suspended load involved collecting samples of stream water and filtering it to measure how much sediment was in suspension. Solution load was measured by chemically analysing the water. The results of the first year's experiment, monthly measurements and the year's total, are shown in table 5.1.

table 5.1.   The Slapton Wood Catchment in Devon. Twelve months denudation.

Month	Runoff (mm)	Suspended sediment yield (kg)	Bed load yield (kg)	Solution (tonnes)
Apr	20	250	0	3
May	10	50	0	2
Jun	9	50	0	2
Jul	9	25	0	1.5
Aug	7	10	0	1.4
Sep	4	10	0	1
Oct	4	10	0	1
Nov	6	25	0	1.2
Dec	9	100	0	1.5
Jan	50	1000	0	7
Feb	110	3500	200	17
Mar	92	2900	380	14
total	333	7930	580	52.6

1   Graph the data given in table 5.1 (Either as four graphs or a combined one. Be careful to consider your vertical scale.)

2   Describe how each item varies during the year.

3   Describe how the three kinds of sediment yield relate to runoff.

4   Try to explain these relationships.

Bed load accounted for only 1 per cent of the year's removal, suspended sediment for 15 per cent and the 'invisible' solution for a massive 84 per cent. (This was potassium, magnesium, calcium, sodium.) Some of the chemical outputs measured would be the result of farmers applying artificial fertilizers. About 4200 kg had been spread on farmland in the basin that year which resulted in excessive nitrates flowing into Slapton Ley. In spite of this human interference, the year's figures remind us that chemical weathering is an important part of landscape stripping.

The monthly pattern in table 5.1 is interesting, as all the bed load, 93 per cent of the suspended load and 72 per cent of the solution load left the basin in January, February and March. These winter months were those when runoff was greatest (fig. 3.15). Streams obviously do much of their work when they have a lot of water in them. In the Slapton stream, suspended sediment concentrations of 3 ppm were found when the stream had a discharge of 2 litres/sec. When the discharge rose to 100 litres/sec the concentration was 60 ppm (don't forget there is more water too, so that the sediment removed is even greater). Bed load movement was even more extreme. Below 42 litres/sec none was found to move into the sediment trap at all. At 50 litres/sec 1 kg arrived in the trap per hour, and at a discharge of 70 litres/sec 4 kg of rock arrived each hour. The figures here apply only to a small Devon basin measured over a year. What is significant is the *relative* size of the solution, suspended and bed loads. These figures confirm a point made in the last section – that the ability of streams to do work increases with floods.

Raindrops hitting the surface (rainsplash) can dislodge finer soil particles and gradually move them downslope where they ultimately end up being carried away by the stream. When the soil's infiltration capacity is exceeded (page 22), overland flow occurs as thin films of water surging spasmodically downslope, when it may pick up small loose particles of soil and deliver them to the streams. However, a vast proportion of the water's energy is used up in overcoming friction as it moves around and across soil particles and vegetation. Where the water becomes concentrated in **rills** the situation is transformed. Energy loss is reduced and it flows faster, picking up (or entraining) soil and rock particles from its bed. In this way it deepens and widens its channel in every storm.

Where a water table intersects the surface, springs and permanent streams are found. In this case the water is really 'in business' as the running water continually erodes its channel, by hydraulic action and the abrasive effect of its load (the processes already mentioned in our discussion of the floodplain meander). In times of flood the ability of the stream to carry load and achieve erosion is transformed. At times of lower flow its bed may be littered with boulders and pebbles stored until the next storm. In some locations pebbles may be swirled around so that they act as 'drill bits' scouring out **potholes** in the river's bed. A particularly large pothole and smoothed river bed is shown in fig. 5.13.

Where is the load from?

1. Slopes

2. Bed and banks

73

The river's erosion in headwater sections is largely vertical. The river could be imagined as a saw cutting into wood. However, there are few river valleys looking like saw cuts, i.e. gorges with steep sides. Weathering and mass movement widen the valley walls whilst the river carries the debris away. When the valley sides are steep it suggests that either the rock is particularly resistant to slope processes and weathering or that the stream is eroding quickly in a vertical direction compared to the slope processes.

Moving down river

So far we have looked at only two sections of the river: the floodplain and the smaller upper tributaries. Two of our earlier examples came from the River Wye, whose basin is shown in fig. 3.7. Tables 5.2 gives details of its bed elevation and drainage area at intervals down its 250 km route from the slope of Plynlimon to the Bristol Channel. In its first 20 km the river falls 422 m. What do you notice about the changes in elevation as we follow the Wye downstream? Its **long profile** would show steeper gradients in the headwaters. Many rivers have this kind of long profile. You might do some map analysis to check this.

The area drained by the Wye is also shown in table 5.2. This increases downstream, but irregularly, The increase between 140 and 160 km, for example, reflects the joining of the Lugg, a major north-bank tributary. As the drainage area increases the discharge of the river also rises.

table 5.2. *The River Wye: bed elevation and drainage area*

Distance downstream (km)	Bed elevation (m)	Drainage area (hundreds of km²)
source	677	—
20	255	1
40	172	4
60	120	11
80	83	14
100	60	17
120	51	18
140	42	19
160	32	31
180	27	32
200	19	33
220	9	40
240	0.5	41
250 mouth	—	42

figure 5.13   Potholes exposed in the bed of the Mistaya River, Alberta. The geomorphologist provides the scale!

Using the data in table 5.2, draw the long profile of the River Wye, using the vertical axis for the bed elevation and the horizontal for distance downstream.

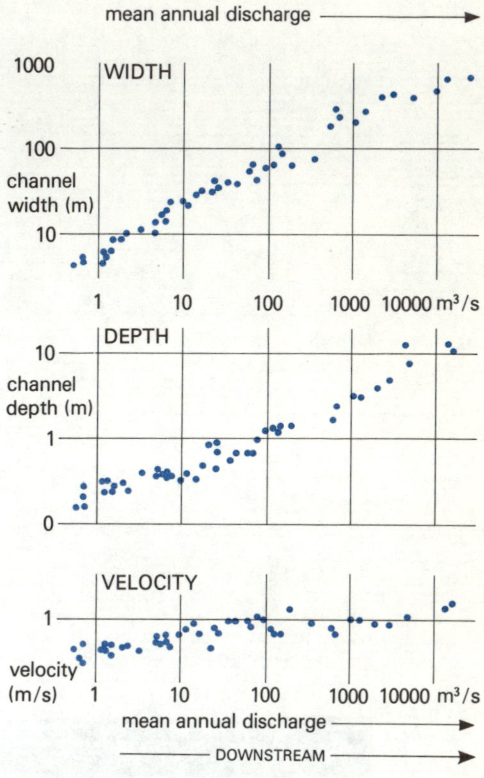

figure 5.14   The Mississippi–Missouri river system. Changes in width, depth and velocity with discharge.

Each dot is a particular point of measurement. You may have noticed the scales. Because of the size of this river system there is a tremendous range of widths, depths and velocities as we journey towards the mouth. The scales are therefore logarithmic. Had the graph been drawn on ordinary arithmetic scales, it would have been far too wide to fit on the page, and it would also have made it more difficult to see trends.

The other river system mentioned in this chapter is the Misissippi–Missouri. It is the most surveyed major river system in the world and fig. 5.14 illustrates some of the results. The horizontal axes are river discharge; in simple terms, they represent a downstream direction as you move to the right. At the top is a scattergraph of channel widths against discharge. The trend of the points tells us that river width increases as discharge increases. The same relationship holds with depth. In other words, *as we move downstream the width and depth of the river increase.*

The bottom graph shows the river's velocity. The fact that it gets faster towards the mouth does not fit the 'image' of a big, slow, majestic river entering the sea contrasted with a fast-flowing mountain tributary. Its gradient, or channel slope becomes more gentle downstream so how is it able to flow faster? The answer is in the *efficiency of the channel.* Its width and depth have increased, so that the proportion of the river's cross section in contact with its banks is less. In the headwaters streams are small, they have a relatively large 'contact area' with their beds and banks. Such contact causes frictional losses and results in slower flow. Downstream this porportion is reduced, so in spite of the lower gradients the mean velocity of the streams is able to increase.

Rivers are agents of erosion, transport and deposition and fig. 5.15 illustrates one of these facets for the lower Mississippi. The diagram shows the relative proportion of fine, medium and coarse material making up the river's load. It also shows the mean size of the bed material. It illustrates how the 'load' changes downstream. This is because the particles being transported by a river rub against the banks, the bed and each other. As a result they become progressively smallers as they move downstream. You may have seen evidence of this in the rounded gravels and pebbles which litter many stream beds. They did not start as rounded fragments. Movement has rubbed them against each other and abraded their sharp edges.

Describe what normally happens to discharge, width, depth, elevation, load and velocity as a river is followed from source to mouth.

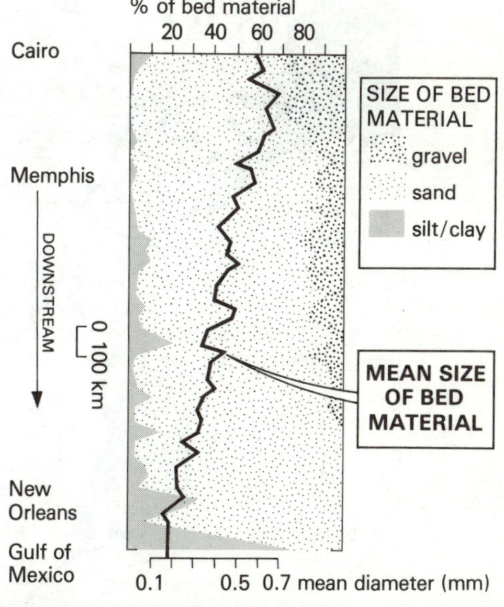

figure 5.15   The lower Mississippi: changes in the size of bed material

The river mouth

The words below were written in 1789 by Alexander Mackenzie, the first European to reach Canada's Arctic coastline after an epic canoe trip down the river which now bears his name. He was travelling through a maze of channels and lakes covering an area 130 km north–south and 40 km wide, bounded by hills to east and west. If you look at fig. 5.16 you can appreciate why Mackenzie was at a loss as to which way to take to reach the Arctic ocean. The detailed mapping of the Mackenzie's mouth awaited the coming of aerial survey after World War II.

'... here the river widens and runs through various channels formed by islands, some little more than banks of mud and sand, while others are covered with a kind of fir. Their banks display a face of solid ice intermixed with veins of black earth and as the heat of the sun melts the ice the trees frequently fall in the river. So various were the channels of the river at this time that we were at a loss which to take.'

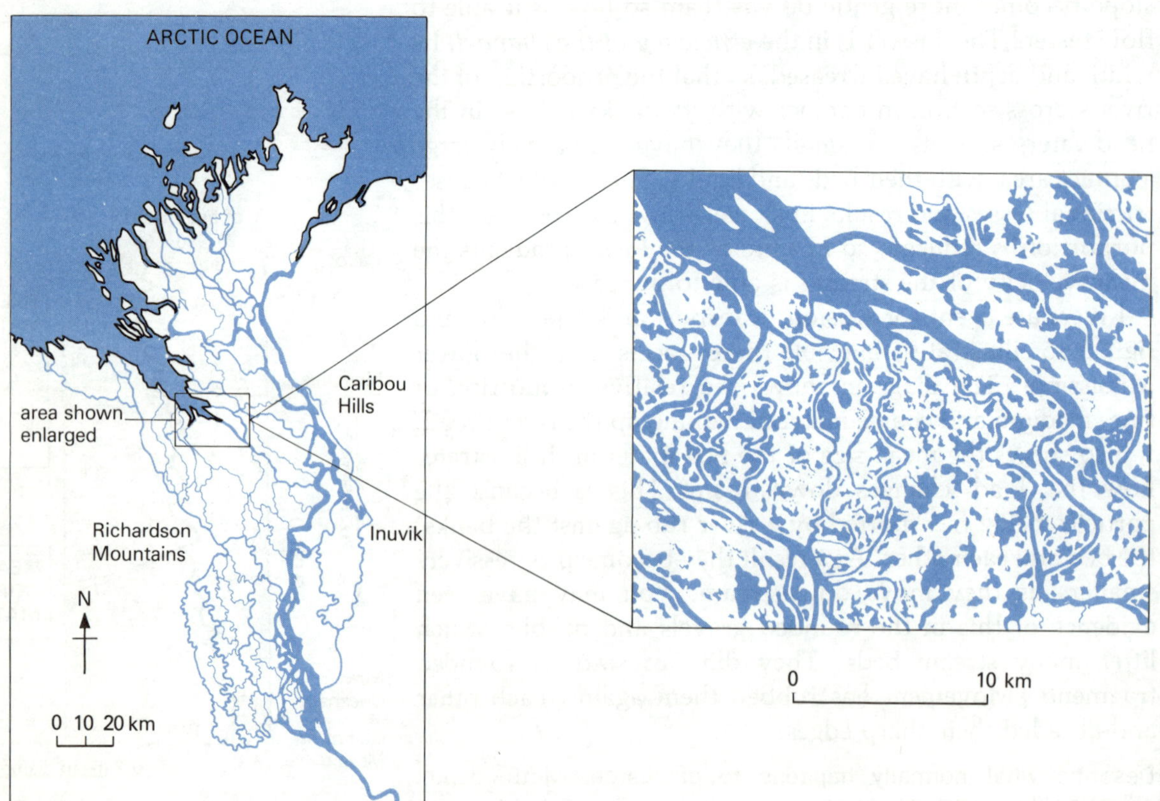

figure 5.16   The delta of the Mackenzie River

The Mackenzie pours water and sediment into the sea. The finer sediment may 'jet' into the sea for tens of kilometres but the heavier material in the river's load may be dropped at the mouth to form a **bar**. Early explorers often encountered such bars, the water was often not deep enough to float a canoe. Such deposition within the channel can lead to the growth of a **shoal**. Water flowing on either side of it splits the channel and leads to the formation of **distributaries**, channels carrying water away from the river. Levees grow along these distributaries and **crevasse splaying**, when floodwaters breach levees (fig. 5.9), can also help to build new land. In the quieter 'backswamp' areas, where plants can slow sediment-laden floodwaters, deposition is also encouraged. The processes at work in such a river mouth environment are summarised in fig. 5.17. (If you look at the enlarged portion of fig. 5.16 you may find evidence of these.)

The Mackenzie carries only about 2 000 000 tonnes of sediment a year into the sea. The Mississippi, draining an area not quite twice as large, carries an impressive 312 000 000 tonnes a year into the Gulf of Mexico. (A figure equivalent to about 100 tonne for each km² of the basin!) The river is also

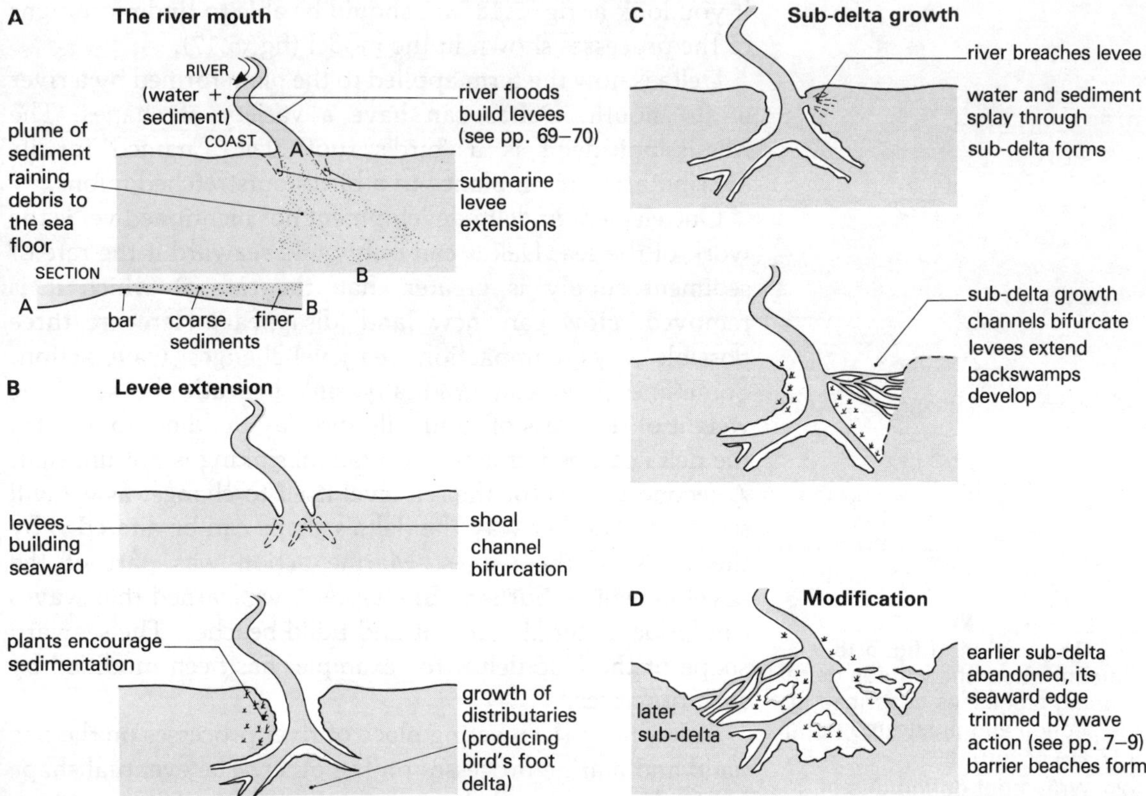

figure 5.17   *A model of delta growth*

figure 5.18   *Part of the Mississippi delta*

carrying material in solution, around 130 000 000 tonnes a year. With such a vast supply of material, it is not surprising that the Mississippi has been able to build an impressive delta. If you look at fig. 5.18 you should be able to find more signs of the processes shown in the model (fig. 5.17).

Delta is now the term applied to the plain formed by a river at its mouth. Deltas can have a variety of shapes. The Mississippi delta is a 'bird's foot' delta, named for its distributaries' resemblance to a bird's outstretched talons.

One element in delta development not mentioned yet is the work of the sea. Deltas can only build seaward if the rate of sediment supply is greater than the rate at which it is removed. How can 'new land' disappear? Here are three possible ways: compaction; sea level changes; wave action. Sometimes land can gradually sink beneath the sea. The weight of the mass of sand, silt and clay building up to form the delta can be immense, so gradual sinking is not unusual. A second way is for the sea level itself to change, as we will see later. Another way the delta's shape can be altered is by the work of the waves. Marine action was part of the development of Surtsey. In chapter 1 we learned that waves can erode material, carry it and build beaches. The seaward shape of the Nile delta, for example, has been modified by such processes.

Deltas lie at the meeting place of river processes on the one hand and marine processes on the other. The eventual shape of the landform will depend on the balance between these two geomorphological machines.

1   On a sketch of fig. 5.18 indicate where the various delta shaping processes described in the section and illustrated in fig. 5.17 are occurring.

2   Write brief definitions of; levee; distributary; bird's foot delta; backswamp.

78

Rivers over time

Rivers are the visible 'veins' of the landscape carrying water to the sea as part of the hydrological cycle. They also act as 'conveyor belts' for the *rock cycle*, carrying material from their beds, banks and surrounding slopes to the sea. Their channels and floodplains also act as stores of sediment. These erosion, transport and deposition processes have been explained earlier in this chapter by using examples of present-day rivers.

When we look at what rivers do over longer periods of time we need a different approach. One way to do this is to *model*, make some assumptions and simplifications and reason how things may have occurred. We could start by imagining a fresh piece of land, of uniform rock hardness, recently emerged from the sea. How could a drainage system evolve? In the beginning the drainage could be disorganised, overland runoff directions changing with every storm. By chance (i.e. randomly) some drainage lines (rills) may emerge which persist between storms. These have an inbuilt advantage: water flows quicker in them and is able to deepen them. A pattern of semi-parallel downslope rills can therefore gradually etch itself into the slope. Unvegetated embankments of motorways during construction can show this pattern of rills. However, we rarely find this parallel pattern in the natural landscape so something must happen to change it.

The 'something' is probably a deeper rill. Once in existence, such a rill deepens, lengthens and widens by a 'positive feedback' situation, i.e. success breeds success. Tapping a bigger contributing area, its discharge, velocity and *erosional energy* are greater. Over time it could therefore *dominate* larger and larger areas of the slope's drainage. It could 'capture' the drainage of the adjacent rills which could then run *across* the slope to the main 'stem' of the emerging 'leaf'.

The pattern of the emerging network may vary in appearance. On relatively uniform rock it may become tree-like, or **dendritic**. Where alternate hard and soft rock outcrops occur it could become rectilinear, grid or *trellis* like (discussed in chapter 8). Initially the **drainage density**, the length of the channel per unit of area (km per km^2), might be small. As the network grows and integrates, it increases. The ultimate density will depend on climate (i.e. the runoff surplus resulting from precipitation inputs to the basins, and their evapotranspiration losses) and on whether the rock is impermeable or not. When streams become organised in this way their long profile might resemble that of the River Wye (table 5.2), with steep headwaters and with gradients declining downstream.

In such a situation the headwater tributaries would be the scene of vertical erosion. The stream's energy, small as it is in uneven and irregular channels, is concentrated on downward cutting. A swinging channel can therefore be imagined as cutting vertically, leaving the bends as spurs which interlock across the main trend of the valley. Downstream, as tributaries join together and the river's size (discharge) increases, its channel becomes more efficient and there is more energy available for erosion. This energy is focused into channel deepening and widening.

We can therefore conjure up a model sequence. Firstly, a gentle plain which has emerged from the sea. Gradually a drainage network evolves and etches itself into the plain. Valleys deepen and widen as the streams erode and transport rock to the sea. The areas between the streams are gradually consumed and the landscape is dominated by slopes. Rivers then begin building new flat surfaces, at a lower elevation than the original surface, their floodplains. Such building is achieved by lateral erosion of the migrating meander belts as well as by the depositional work summarised in fig. 5.8. The rivers carry rock to the sea so the landscape slowly lowers. Hills, final fragments of the original surface, become lowered and a low relief surface (a **peneplain**) dominates, with a few residual hills (called monadnocks).

Such a sequence would take many millions of years to achieve. It was first described by a pioneer geomorphologist, W. M. Davis, who termed the model the *cycle of erosion* ('cycle', as it both began and ended with a relatively flat surface). Davis divided the cycle into three stages: *youth* (when the landscape was being etched and relative relief increased), *maturity* and *old age* (the peneplain stage). The terms were also applied to rivers, youthful being the stage of vertical downcutting, maturity one of valley widening and meander development and old age being the lower courses of rivers where deposition was dominant.

This cycle idea is interesting, but difficult to prove! Three-quarters of a century after Davis our understanding of landscape processes has increased, but not enough to envisage the detail of development over time. The words themselves are sometimes unhelpful, e.g. 'old age' applied to the lower course of a river. As we have discovered already, the river here is flowing faster and able to change landforms quickly.

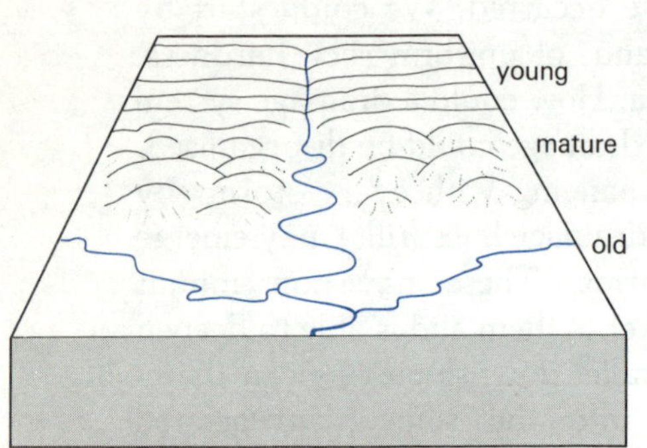

figure 5.19   The W.M. Davis model of fluvial landscapes

Youth, maturity and old age

Environmental change and rivers: the effects of man, climate and sea level

One agent of environment change in recent times has been *people*. Figure 5.20 shows changing river channel conditions as an area was cleared, farmed and urbanised.

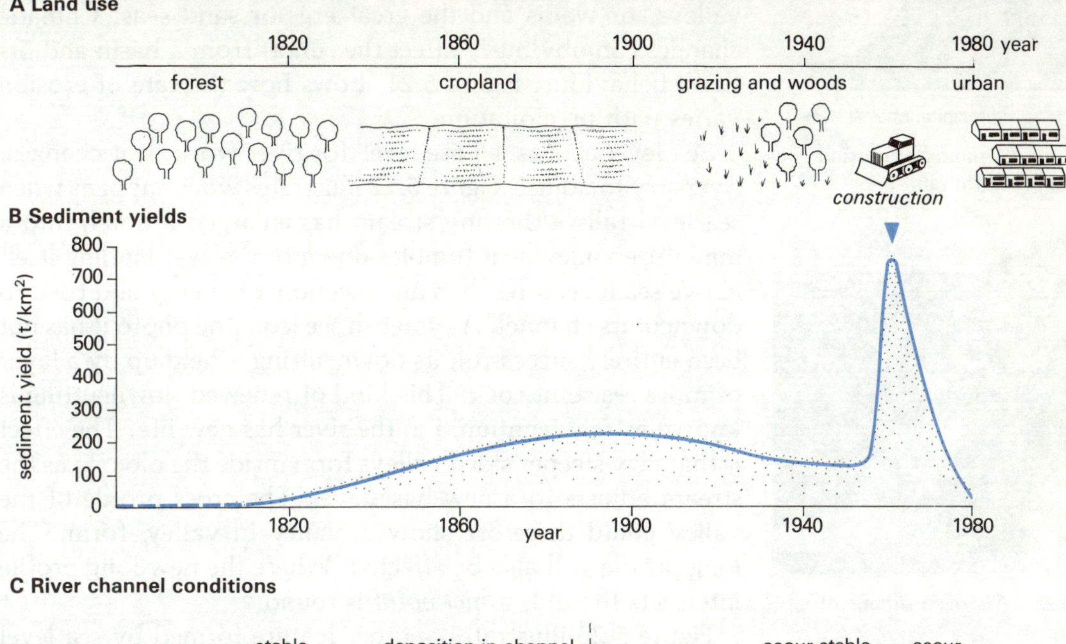

figure 5.20   Wolman's model of changing sediment yields. Based on the history of the Piedmont region of the Mid Atlantic states in the USA.

Controlling rivers to prevent flooding can also have effects. The Missouri provides us with an example. In its natural state it earned its nickname of 'Old Misery' with its behaviour (page 65). It is hardly that today. It has been transformed by seven dams into a virtual chain of lakes. These store an amount of water equivalent to three years' runoff! This is released in a controlled way to generate electricity and to restrain the spring floods. The quieter waters of the reservoirs became huge sediment traps. As a result the water flowing downstream away from the dams is clearer and less charged with sediment. It has therefore been able to erode its bed. Measurements at the Garrison Dam have shown a downcutting of 25 cm a decade! Such effects of damming are most noticeable with large rivers flowing through semi-arid or desert areas, with few or no tributaries adding water and sediment to them, e.g. the Colorado and Nile.

Examine fig. 5.20 (you will also need to remember the basin hydrological cycle from pages 20–23).

1   Describe what happened between 1820 and 1860.

2   Explain these changes.

3   What happened between 1900 and 1950?

4   Explain why sediment yields increased during construction.

5   Can you suggest some reasons why sediment yield from urban areas is only about a quarter of that from cropland?

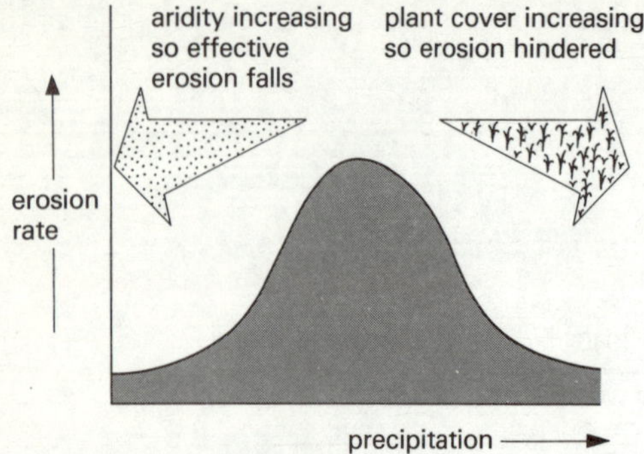

figure 5.21   A model of erosion rates and precipitation

figure 5.22   A stream downcutting

figure 5.23   A real life hardware model! The beach sands here have been cut into by the tiny stream. As the tide fell the rivulet tried to cut itself to the ever-lowering 'base level' of the sea. As it did so it left remnants of its earlier 'valleys' perched as miniature river terraces.

Another environmental change can be *climatic*. An area may become hotter or cooler, wetter or drier. Today the central Sahara has a low and erratic precipitation. A mere 20 000 years ago, in a wetter period, runoff would have been dissecting the highlands and depositing sediment on the lower ground. The remnants of these two processes are the dry valleys, or **wadis** and the great **ergs** or sand seas. Climatic changes can obviously affect the runoff from a basin and, its flood behaviour. Figure 5.21 shows how the rate of erosion varies with precipitation.

Sea level acts as a 'base level' for river work. If it changes, rivers try to adjust. Figure 5.22 illustrates what happens when sea level falls – the tiny stream has enlarged a notch into a miniature valley as it tumbles down to the sea. Finding itself above sea level it has had an 'injection' of energy and tried to downcut its channel. As you can see from the photo it has not been entirely successful, its downcutting is held up by a layer of more resistant rock. This kind of renewed downcutting is known as **rejuvenation**, i.e. the river has new life. The effect is that new steeper sided valleys form inside the old one as the stream adjusts to a new base level. The cross profile of the valley could therefore show a 'valley in valley' form. The long profile will also be affected. Where the new long profile intersects the old, a *nickpoint* is found.

Figure 5.23 illustrates another feature formed by sea level change. You can see that the tiny rivulet's downcutting has left remnants of its earlier bed as miniature **terraces**. At larger scales, and over very much longer time periods, rivers may produce such features. Their terraces can represent earlier valley floors (floodplains) now left abandoned as the river downcuts to a new base level. River terraces can be useful for man. Many towns and villages are sited on them as they provide dry-point or flood-free situations. The level surfaces of terraces and their easily worked soils may also provide the focus for farming.

Summary

- Runoff from the land is part of the cycling of water.
- Rivers erode their beds and banks.
- Rivers transport rock material from their beds and banks and also from their surrounding slopes. The material is removed in solution, in suspension, by saltation and by dragging.
- Rivers deposit material when their discharge and velocity fall.

- Erosion, transport and deposition act together to produce landforms like meanders, floodplains and deltas.

Exercises

1 Examine fig. 5.24A, which shows the Rhône delta.
(a) What evidence is there for the kind of environmental processes mentioned in the chapter?
(b) Describe the ways in which people use the delta environment.

2 Examine fig. 5.24B, which shows the Nile delta.
(a) Describe and explain the shape of this delta.
(b) The annotation mentions some of the changes in the delta since the completion of the Aswan dam and the filling of Lake Nasser behind it. Briefly explain why these have occurred.

3 Figure 5.25 shows some evidence of channel changes on a Welsh river.
(a) Describe and explain the main changes between 1841 and 1971.
(b) If you were an engineer responsible for planning the future maintenance of the railway embankment over the next 100 years, where would you expect problems to occur? Give reasons for your choice.

4 Your town or suburb may well have a river flowing through it. Using O.S. maps, Water Authority data, old newspaper reports and other references from your local library (and possibly interviews with long-established local people) produce a report to;
(a) illustrate the distribution of the floodplain;
(b) describe the land use on it;
(c) explain why the river floods;
(d) illustrate the effects of flooding on people and land use;
(e) explain what is being done about it.

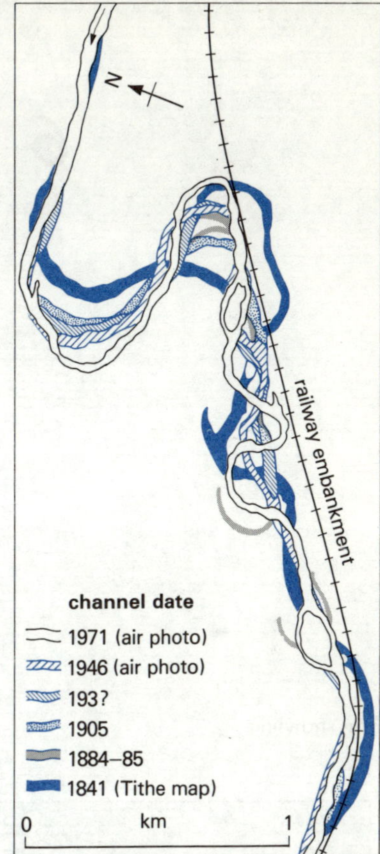

figure 5.25   The River Tywi, south-west Wales

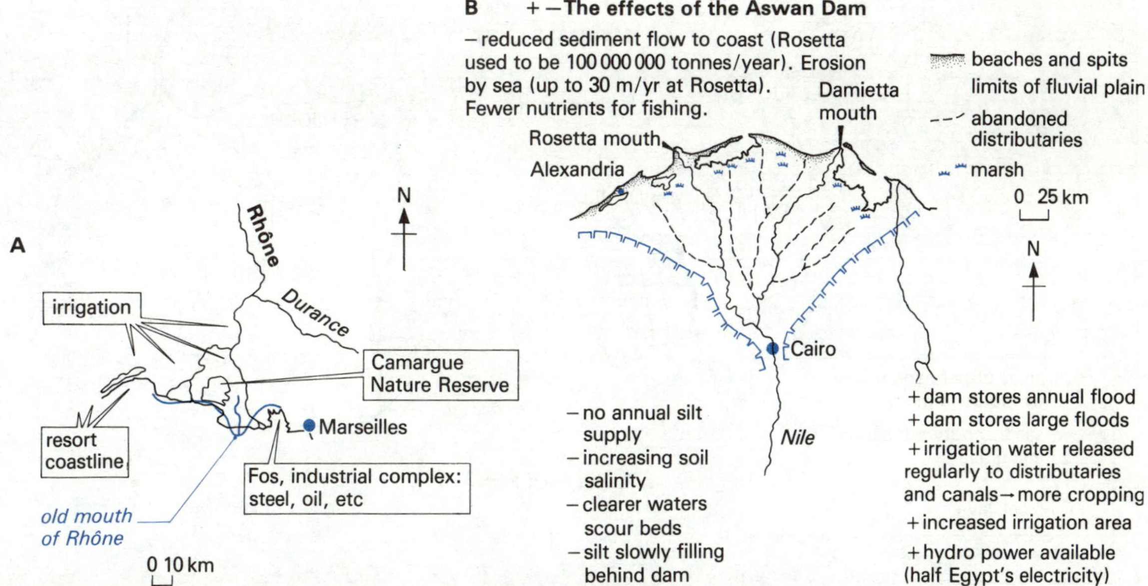

figure 5.24   *The Rhône and Nile deltas: environmental changes and human uses*

6 Ice

Introduction: Myrdalsjökull

Just 75 km to the north-east of Surtsey is the area shown in fig. 6.1. Most of the land above 800 m is permanently covered with snow and ice. Such areas are called Jökull in Icelandic. In spite of Iceland's northerly latitude this lowland coastal strip has winter temperatures hovering around 0°C. This reflects the warming influence of the North Atlantic Drift and on-shore winds. Inland, altitude increases and temperatures fall. Such cooling with height means that more precipitation falls as snow. It also causes the period of summer melting to become shorter. Eventually a balance is reached when all the winter's snow cannot be melted. This is known as the **snowline**.

The snowline

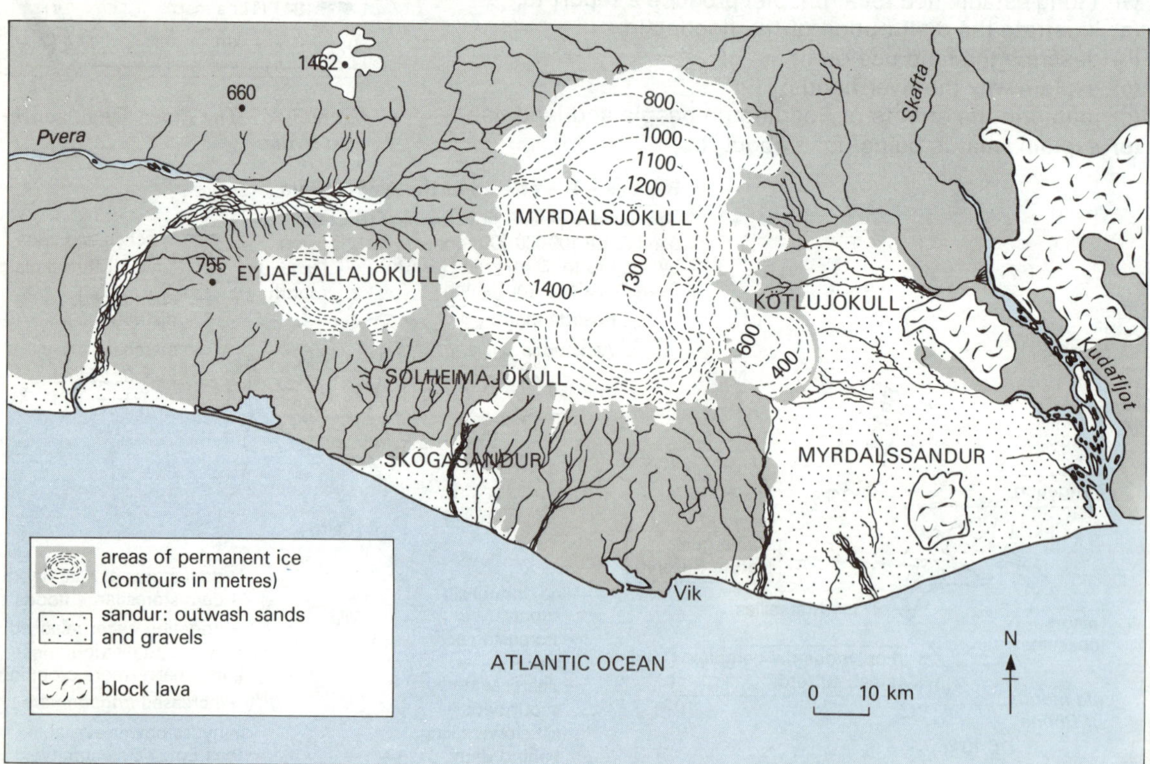

figure 6.1   *Myrdalsjökull, Iceland. The location of this icecap is shown in fig. 1.1.*

Most people in the UK see snow for only a few days each year, when it falls as light flakes. Its density is low, 10 mm of snow may only be equivalent to 1 mm of rain. When it accumulates over the years it does not stay like this. It becomes transformed into **firn** and ultimately into *ice* by the weight of later snowfalls and partial melting and refreezing. If we look at Solheimajökull on the south-west of Myrdalsjökull and the larger Kötlujökull on the south-east, we can see what happens to this accumulating snow and ice. In both places the ice extends to below 200 m, i.e. well below the snowline. The ice accumulating on Myrdalsjökull is 'escaping' downwards. At these lower heights it thaws and meltwater streams can be seen flowing away from the edge of the ice. As it melts it is replaced by ice continually moving downwards from the higher zones where melting was less than snowfall.

Ice and meltwater are potent agents of landscape change – eroding, transporting and depositing rock material. On the map the streams flowing from the Kötlujökull do not have permanent positions. They are flowing onto a huge spread of sand and gravel, known in Iceland as a sandur. The melt-waters have carried and deposited rock material eroded earlier by the moving ice. Such deposits are known as **outwash** plains. Trying to maintain roads across them is difficult because of the massive meltwater floods in spring. In some places away from the ice front the sand is so fine that when it is driven by the wind it can obscure the sea edge in a yellow mist, one factor which makes this coast dangerous.

The distribution of ice today and in the past

Myrdalsjökull is only one of Iceland's ice covered areas. Figure 6.2A shows the other areas of permanent ice in the northern hemisphere, dominated by the huge Greenland Ice Cap. The smaller ice-covered areas in the middle latitudes occur in the mountains, especially those with heavy snowfall. Thus you can see the influence of two factors in the distribution of ice; 'coldness' (a reflection of latitude and altitude) together with 'snowiness'.

During the Pleistocene period (very recently in geological terms) ice covered much larger areas, as can be seen from the map. Locking all this water up in such huge 'stores' caused sea level to fall to levels 200 m below those of today. This climatic and sea level change had repercussions on river behaviour elsewhere in the world. It also helped mankind to move from place to place, as you can imagine from looking at the coastlines at the ice maximum.

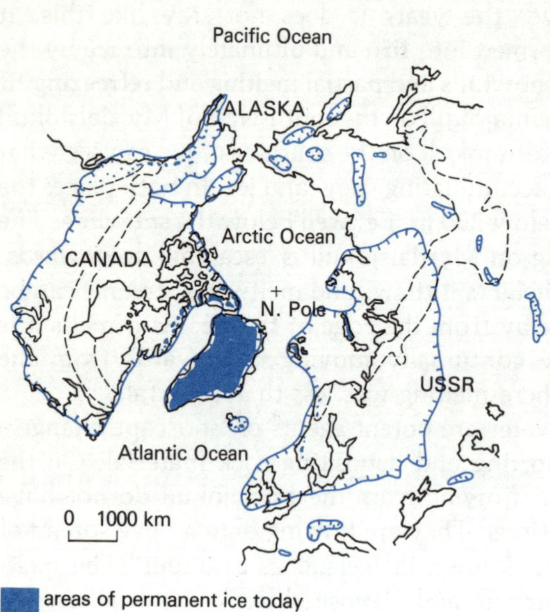

areas of permanent ice today
limits of the Pleistocene ice sheets
limits of permafrost (permanently frozen ground)
today's shorelines

figure 6.2   A.Present and past ice distributions in the northern hemisphere. It is important to remember that the Antarctic ice cap contains the vast majority of the world's frozen water today.
B. Ice Age Britain

1   On a copy of fig. 6.2A mark in the names and heights of the mountain ranges where ice is found today.

2   Annotate your map to show how latitude (northness), altitude and high precipitation (snowiness) combine to produce today's permanent ice.

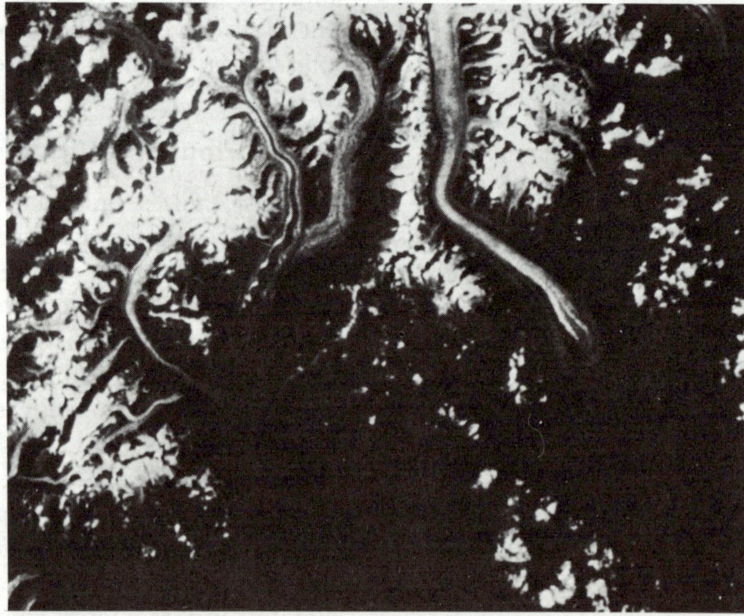

figure 6.3   Glaciers in the Alaska Range around Mount Mckinley (6194 m). The glaciers appear as light grey tongues moving down valleys from the permanent snowfields (white). Their surfaces are bordered and striped by moraines.

Mountain glaciation

Myrdalsjökull is a small **ice cap** totally covering the underlying rock. In mountain areas where peaks protrude above the snowfields we have a different situation. The moving ice is confined by the valley walls.

Cirque glaciers

The smallest glaciers, found on the flanks of mountains, are known as **cirque** (or **corrie**) **glaciers**. These begin as snow patches, around and beneath which processes like freeze-thaw rock destruction occur. As the snow and ice thickens, it begins to move slowly downslope under the influence of gravity. You can see from fig. 6.4 that the ice is 'rotating' in its hollow.

The moving ice is not just slowly moving frozen water from the accumulation zone to the melting zone – it is also acting as an agent of landscape change. Ice itself is relatively

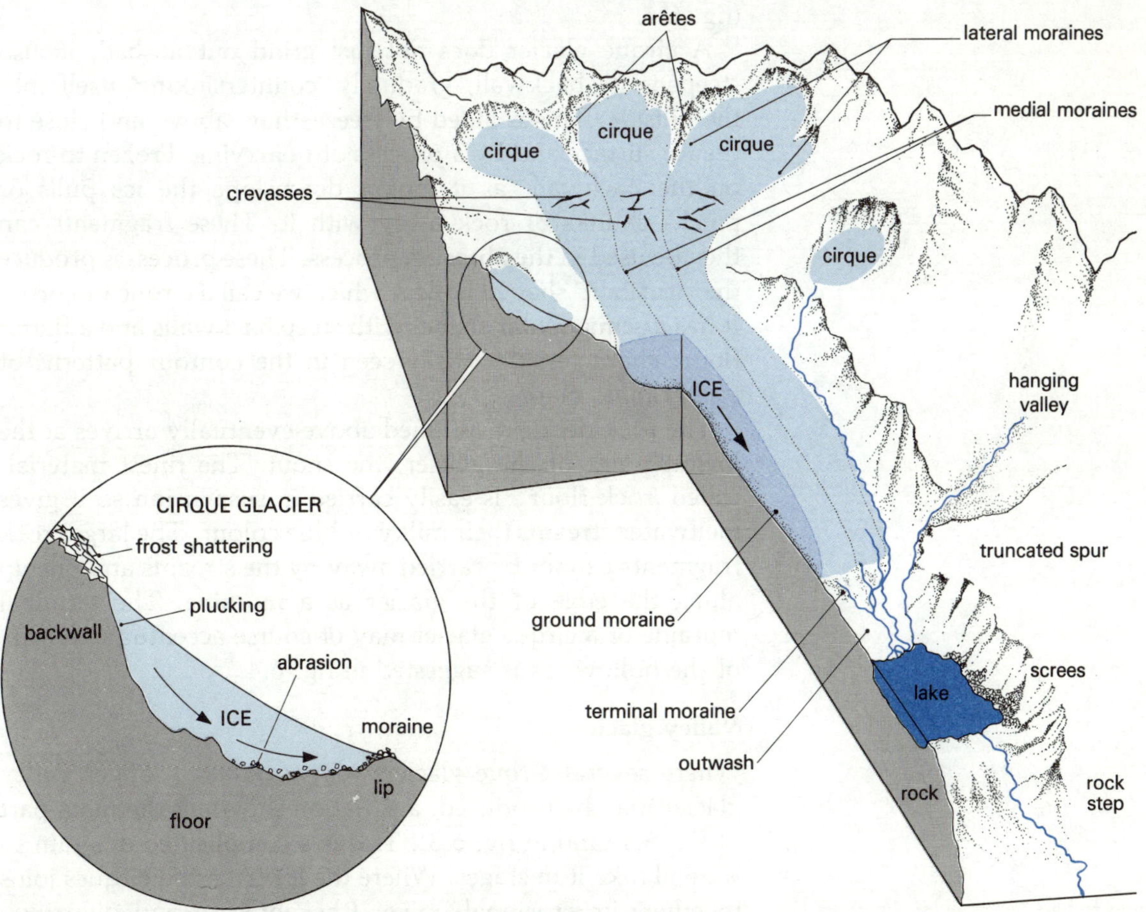

figure 6.4   Mountain glaciation, processes and landforms

'soft' compared to rock and to envisage it eroding is like imagining a heavy block of clean soap scratching a plastic laminate surface. But glacier ice may contain rock particles and this transforms the situation. It is like dipping our soap into sand and then rubbing it across the plastic surface. The sand grains act as abrasive tools scratching the surface.

One source of such abrasive tools in the ice of the glacier is *freeze-thaw* weathering. The rock slopes around the glacier are subject to such weathering and resulting debris falls onto the ice where it can be buried by later snowfalls. Embedded in the moving ice, the rock debris scratches the floor and side of the hollow. In the process the debris gets ground into finer material. This process of scouring, **abrasion** or grinding can lower the rock floor of the cirque. Obviously where the ice is thinner and moving more slowly, i.e. towards its lower edge or *snout*, the erosion may be less powerful. This means that the shape of the rock floor of the cirque may become hollowed out, or overdeepened, with a 'lip' at its lower part (fig. 6.4).

A cirque glacier does not just grind out its bed, it also steepens its backwall, gradually 'countersinking' itself into the hillside. This is aided by freeze–thaw above and close to the ice surface and by a process of **quarrying**. Frozen to rock on the backwall, as it moves downslope the ice pulls or plucks chunks of rock away with it. These fragments can then be used in the abrasion process. These processes produce the 'armchair' shaped hollow which we call a cirque or corrie. It has a semicircular shape, with steep backwalls and a flatter floor: characteristics easily seen in the contour patterns of topographic maps.

The rock debris mentioned above eventually arrives at the lowest point of the glacier, the snout. The finest material, called 'rock flour', is easily carried in suspension so it gives meltwater streams their milky or blue colour. The larger rock fragments cannot be carried away by the streams and pile up along the edge of the glacier as a **moraine**. The terminal moraine of a cirque glacier may of course accentuate the 'lip' of the hollow, as is suggested in fig. 6.4.

Valley glaciers

Where several cirque glaciers join together a longer valley glacier may be produced, a situation shown in the main part of fig. 6.4 (and in fig. 6.3). This is a complicated diagram so we will take it in stages. Where the ice from the cirques joins together, its erosional power (the abrasion and quarrying mentioned above) is increased. This means that the main valley can be deepened. Where the ice moves over this steeper slope it cracks and fissures, producing a series of **crevasses** or ice falls (fig. 6.4). When the ice melts it exposes a steep end to the main valley, a **trough end**, above which are perched the cirques.

Glaciers move much more slowly than the streams which carved the original valleys – tens of metres in a year rather than in a minute. This means the cross-sectional area of the 'river of ice' is huge compared to a quickly flowing stream carrying the same amount of 'water' downslope. The ice has a greater contact area with the valley's side and bed. The processes of rock removal widen and deepen the glacial valley. During a glacial period the previous river valley is therefore transformed from a V-shape to a broadly *U-shaped cross profile*, sometimes referred to as a glacial **trough**.

The size of the moving tongue of ice and its inability to turn sharp corners means that the valley glacier tends to straighten its valley. The interlocking spurs of preglacial upland rivers therefore get planed off. After the glacial period these form **truncated spurs**. If the tributary valley was not deepened by ice this can mean that the postglacial stream falls abruptly into the main glacial trough – a landform known as a **hanging valley**

In a valley glacier system there are several types of moraine. Freeze–thaw weathering on the rock walls surrounding the glacier can 'rain' debris down onto the ice. Here it piles up and is moved away forming a **lateral moraine** (figs. 6.4 and 6.5).

Where cirques develop side by side, the rock ridges separating them can become steep-sided frost-fretted ridges

figure 6.5   A lateral moraine, piles of frost shattered rock debris along the side of a moving glacier in the Yukon

figure 6.6   A glacial valley in the Canadian Rockies showing the trough-like main valley

figure 6.7   A view in Glacier National Park

Produce brief definitions of the following, using diagrams if you wish: arête; medial moraine; terminal moraine; lateral moraine; glacial trough; truncated spurs; hanging valleys.

known as **arêtes**. Where two or more cirque or valley glaciers join together their lateral moraines are carried down the main trunk glacier as a **medial moraine**, a belt of rock material forming a line down the glacier (figs. 6.3 and 6.4).

All the rock debris, carried on the edges, on the middle, or within and beneath the ice after it gets buried by later snowfalls or ice movements will be dumped at the valley glacier's snout. Meltwater streams carry only a portion of this away so the rest will accumulate as an end or **terminal moraine**. If the edge of the glacier remains at the same place for some time such a moraine can become quite large. When the margin recedes it will be left as a cross-valley ridge (fig. 6.7).

On, in and around the edges of the lower part of the glacier meltwater streams occur, especially during the summer. These streams carry rock material and deposit it as spreads of outwash sands and gravels (fig. 6.6).

Measuring glaciers

Glaciers are rather like long-term weather records, expanding when it is snowier or colder and shrinking when it is warmer or drier. Trying to untangle their stories is not easy. In 1980, for example, scientists from Environment Canada tried to drill into the ice on Mount Logan. Their report is printed on the next page.

from '*Information North*'

Fortunately not all studies of glaciers are as difficult as this
example. Scientists are usually trying to find out a number of
things, e.g. the volume of ice, its pattern of movement, its
overall budget and the processes at work within and beneath
the glaciers. The **budget** is fairly central to this work. It
involves measuring how much snow falls (the input or
accumulation) and how much water is lost by melting and
evaporation (**ablation**).

After the ice: the landforms produced by mountain glaciation

Figure 6.8 is a map of the Rockies close to the Montana/
Canada boundary. The area now has only a few small cirque
glaciers, tiny remnants of the ice which occupied the valleys
until about 7000 years ago. If you look at the Mokowanis
valley the pattern of the contours and lakes illustrates a
typical collection of mountain glaciation landforms.

At the head of the valley there are five cirques, three still
occupied by ice. The steep headwalls, the semicircular shape

91

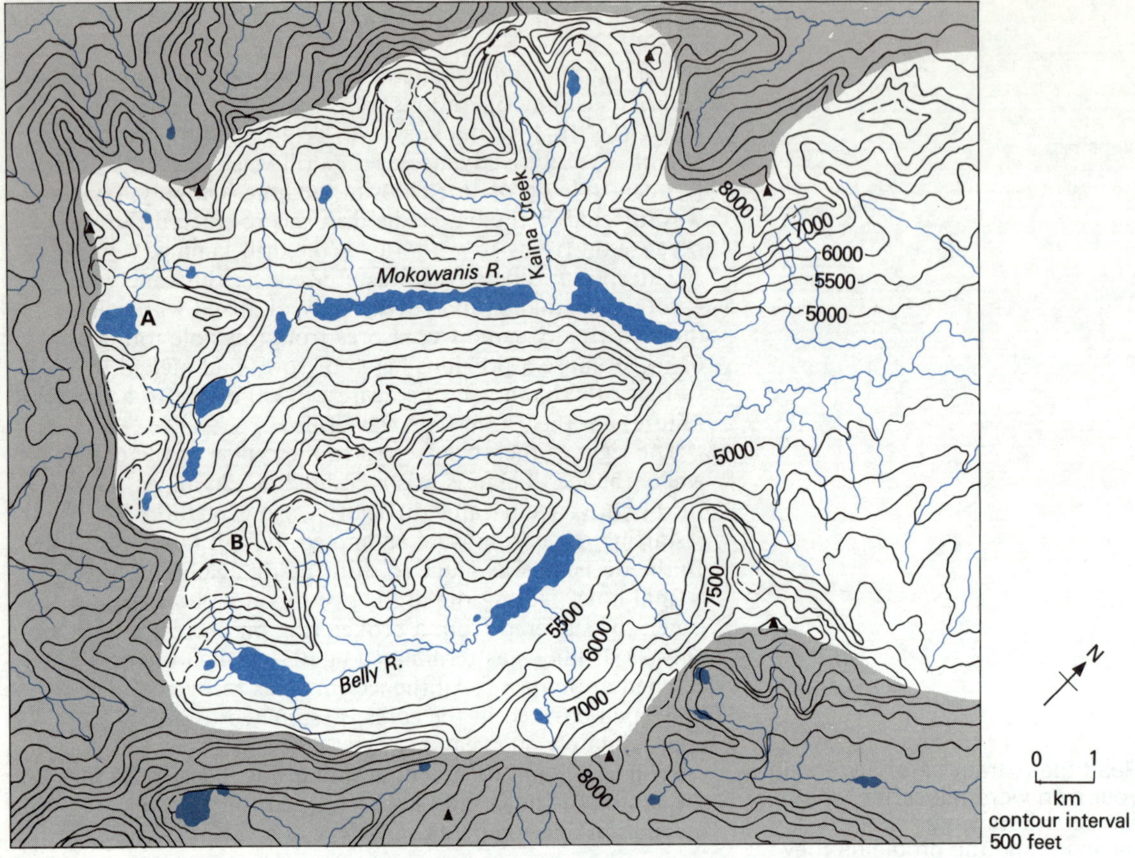

figure 6.8   Glacier National Park, Montana

and flattish floors of the two ice-free cirques can easily be seen from the contour patterns. In one of them (A) you can see a lake, its water held back by a lip of bedrock and moraine. Between the closely spaced cirques are steep ridges, arêtes. Where cirques have carved themselves into the flanks of a mountain, a steep horn or **pyramidal peak** is found (B).

The main part of the Mokowanis valley is almost straight. Notice that its cross profile is the classic U-shape. Its long profile is different from a river valley too. There are steep 'trough ends' where the five tributary cirque glaciers joined together. The floor of the main valley is occupied by two elongated lakes, known as **ribbon lakes**. We cannot tell from the map if they are filling a rock basin or whether they have been ponded up behind a dam formed by terminal moraine. There is some evidence of postglacial change; notice that the Kaina Creek flowing into the valley from the north side has built a fan of deposits (i.e. like a delta) across the valley, splitting the ribbon lake into two halves.

Look at fig. 6.8 and the Belly River area (to the south of the Mokowanis valley discussed earlier).
(a) Produce a sketch map of the valley.
(b) Label all the glacial features you can identify.
(c) Explain how each of them was formed.
(d) Identify any signs of postglacial landscape processes.

In Britain today the snowline is usually a few hundred metres above the summits of our highest hills. Consequently only after heavy winter snow and cool weather in spring do snow patches linger through the summer. As we saw in fig. 6.2B the situation was very different in the recent geological past. The effects of this glaciation can be seen in many highland areas of Britain. The valleys may not be as deep as the Mokowanis, as can be seen from fig. 6.9, but their formation was the same.

Cirques, troughs and truncated spurs are large features of glacial erosion. There are many small-scale features visible after the ice has receded. The smallest are **striations**, scratch marks caused by ice containing moraine moving across bedrock (fig. 6.10). *Quarried* surfaces are another feature where the ice has plucked out blocks of rock. One landform which has been both striated and quarried is the **roche moutonnée**.

Deposition by ice and meltwaters must not be forgotten. Outwash deposits and terminal moraines have been described earlier. Often at the end of the glacial phase the situation becomes quite confused. Stagnating and melting ice dump moraine in all sorts of shapes and sizes. Meltwater strives to remove, rework and resort these deposits. One feature of glacial deposition is the so-called ground moraine carried and deposited at the base of the ice. In some locations this may be moulded and shaped by moving ice. A landform produced by this process is the **drumlin**. These are elongated hummocks, typically a few tens of metres high and hundreds of metres long. The shape resembles an upturned spoon. They frequently occur in swarms, when the terrain is known as 'basket of eggs' topography.

figure 6.9   A glaciated valley in Scotland

figure 6.10   Striations, small scratches in bedrock caused by rock particles embedded in the bed of a moving glacier. Such abrasion produces rock flour, tiny mineral particles. When picked up by meltwater such 'flour' turns streams milky white or bright blue.

Look at fig. 6.2A and describe the extent of the last glaciation.

Ice sheets and continental-scale glaciation

The moving ice which shaped the Mokowanis valley described above was confined between the valley walls. Where ice totally covers an area, an ice cap or ice sheet occurs. This also erodes, transports and deposits. As the map of past ice showed (fig. 6.2A), the ice sheets of the last glaciation would dwarf the Myrdalsjökull (fig. 6.1).

The *erosional* effects of such continental-sized ice sheets included removing the preglacial soil and weathered rock. Where the bedrock was weaker, along joints for example, the ice could quarry out basins. Where it was more resistant rock knobs would be abraded and smoothed. After the decay of the ice sheet the basins would become lakes. Figure 6.11 is a photo of part of the Canadian Shield showing this type of scenery.

The *depositional* effects of the ice sheets include the range of morainic and meltwater features mentioned earlier. The preglacial relief may be totally obscured by glacial deposits (often called boulder clay in the UK). In Norfolk such deposits are up to 10 m thick. End moraines may form low ridges across the country and drumlin swarms occur.

Water is a vital part of the environment around the margins of the ice sheets. Where meltwater flowed beneath the ice it was able to carry and deposit sands and gravels. When the ice finally melted the line of such drainage may be seen as an **esker**, a ridge of sand and gravel (fig. 6.12).

figure 6.11   *The Canadian Shield, a landscape littered with lake basins eroded by the continental ice or ponded back by the irregular deposition of moraines. The ice here was several kilometres thick, only finally melting away about 7000 years ago.*

figure 6.12   *Esker deposits*

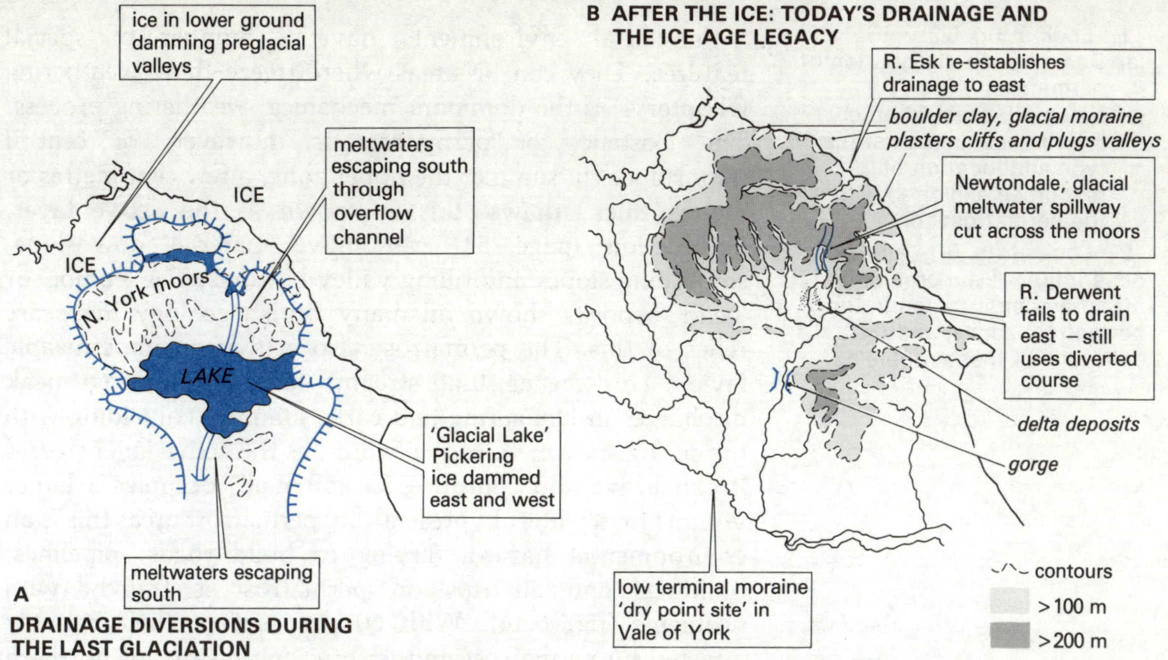

figure 6.13   *Glacial effects in North Yorkshire*

Beyond the ice margin such sand and gravel would have spread out as an outwash plain. The drainage of meltwater around ice sheets can also produce major changes. Figure 6.13 shows the effects of meltwaters in the North York Moors and Vale of Pickering area.

Ice and man

Ice is, and has been, of major importance in shaping the detail of landscape in the middle latitudes. This has been most dramatic in mountain areas like Snowdonia or the Rockies and more subtle in lowlands like East Anglia or the North German Plain. Glaciated mountain areas are a *resource*. Their physical geography, steep, rocky, ice-fretted slopes and lakes provide scenery which attracts winter and summer tourists. The same physical geography makes them attractive as locations for power schemes (hydro-electric and pumped storage) and water supply.

Underground ice

The areas close to the ice margins today and in the past are known as the **periglacial** areas. One way to define these is by the existence of permanently frozen ground, or **permafrost**. The extent of present-day permafrost in the USSR, Canada and Alaska is shown in fig. 6.2A.

1   Describe the erosional and depositional features produced by continental glaciation.

2   Look at fig. 6.13. Write an explanation of how ice and meltwaters have affected the scenery of this area of Yorkshire.

Periglacial environments have a number of special features. They can be areas where freeze–thaw weathering (chapter 4) is the dominant mechanical weathering process. The existence of permafrost is, however, of central importance. In summer the upper zone, a few centimetres or metres thick, thaws; this is known as the **active layer**. Solifluction (page 54) can move material downslope, smoothing slopes and filling valley bottoms. The 'combe' or 'head' deposits shown on many British geology maps are relics of this. The permafrost also acts as an impermeable layer. This means that streams have pronounced peak discharges in late spring and early summer, coinciding with the spring snowmelt. In the winter, as the active layer freezes it can heave and contort – ice, after all, occupies a larger volume than water. In present-day permafrost areas this is an environmental hazard. Trying to build roads, pipelines, buildings and airstrips on permafrost is fraught with problems (fig. 6.14). With 60 per cent of their country affected by permafrost and with a long history of northern development, the Russians pioneered engineering and constructional techniques in such areas.

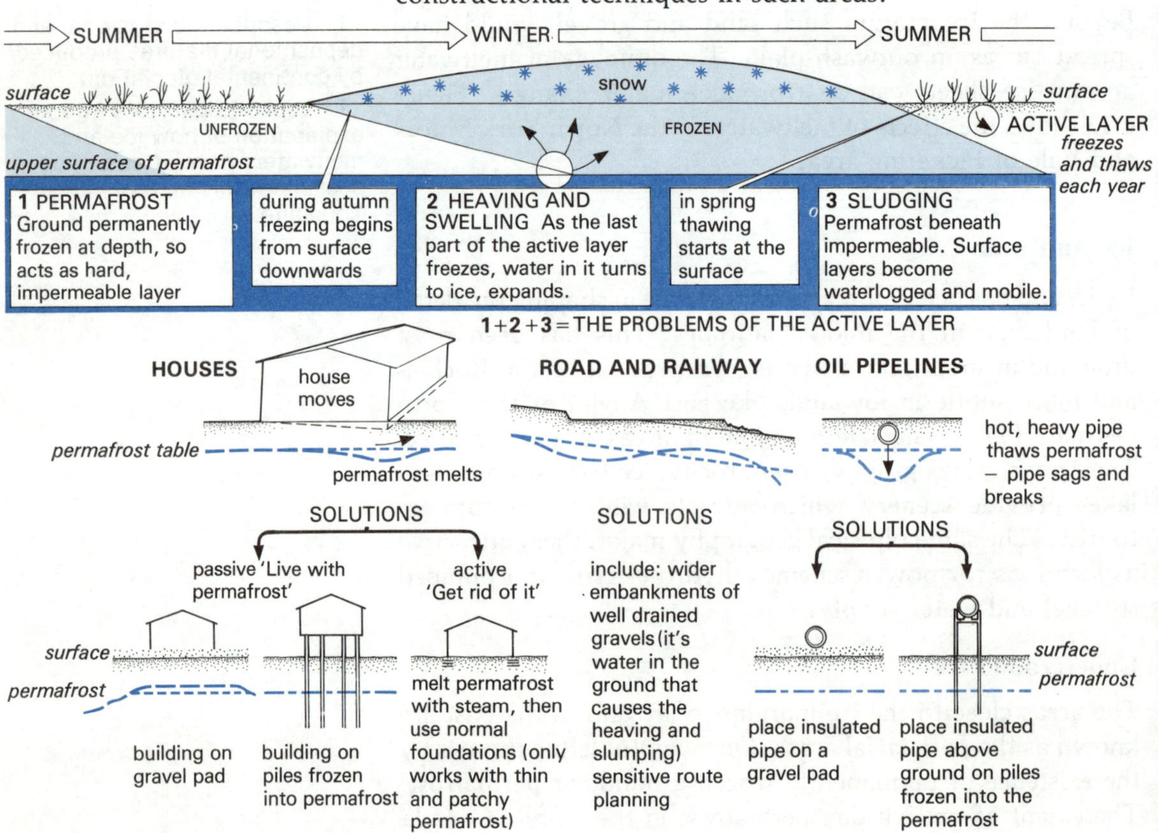

figure 6.14   Permafrost construction

Conclusion

The glaciations of the recent geological past, the Pleistocene period, are a reflection of global climatic change. We are living in an 'interglacial' period. When the ice returns to the middle latitudes our descendants will have to make some drastic adjustments as sea levels fall and farms and cities become covered with ice. *Homo sapiens* are fairly resourceful and should have a few thousand years to adjust! What it does suggest is a final point – that understanding glaciers helps us to understand the history of climatic change and may give clues to the future.

Summary

- Ice is a geomophological machine able to erode, transport and deposit material.
- Valley glaciers modify mountain valleys by their erosion and deposition. They create scenery which is a tourist resource.
- Continental ice in the most recent geological past affected huge areas in the middle latitudes. It has left a legacy of erosional (shields with their rock knobs and lake basins) and depositional features (like moraine-veneered plains).
- Permafrost today underlies large areas, especially in the USSR and Canada. It affects streamflows and mass movement. It makes the exploitation of resources more costly.

7 Coastal environments

Introduction: processes and pressures

Part of the Dorset coast is shown in fig. 7.1. In the north is the urban area of Weymouth. It began as a small port at the mouth of the River Wey, exploiting the resources of the sea through fishing and trade. Quarrying limestone on Portland was also aided by the sea, as it provided the only way to carry stone in the pre-railway era. The sea is also a factor in the more recent growth of the urban area. The sheltered bay in the lee of Portland became a naval anchorage and this military function continues today. The second marine resource was the curving sandy beach in Weymouth Bay.

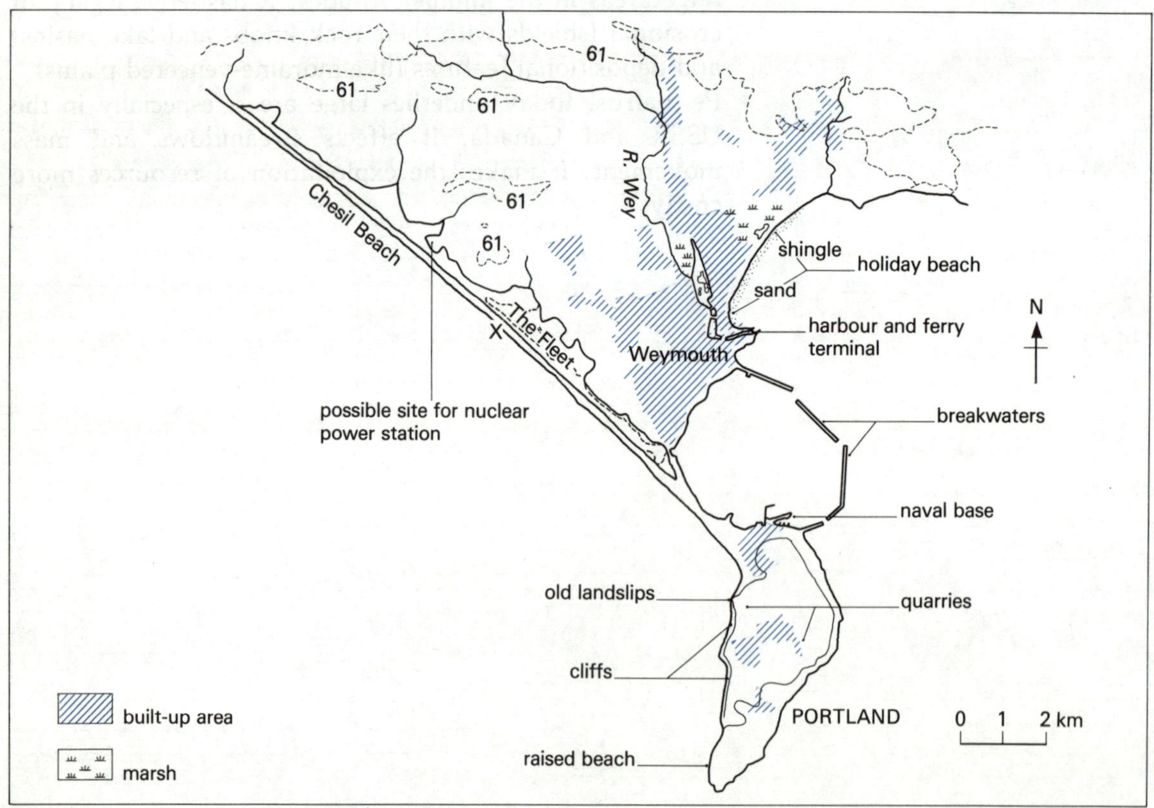

figure 7.1   *The Weymouth area. This section of the Dorset coast, used for a variety of purposes, is subject to a range of natural and human pressures.*

During the nineteenth century, as seaside holidays became fashionable and as railways allowed people to travel from inland industrial cities, so the holiday industry began. This coast has therefore been exploited for transport, fishing, recreation and defence.

Some coastal uses do not mix easily, such as industry and tourism. The Weymouth area has no large-scale industry but one potential conflict emerged in the early 1980s when a site at Herbury was considered for the location of a nuclear power station.

The coastal environment, the inlets, anchorage, beaches and rocks, which man has been using over the last 1000 years are not unchanging. Some changes occur over a few hours and irregularly. Winter storms for example, with strong westerly winds, can produce large waves which break on Chesil Beach. At its Portland end this beach is made of large pebbles or cobbles. Powerful storm waves can throw these over the top of the beach ridge, so that the road to Portland is blocked and buildings damaged. The regular cycle of the tides, in contrast, alters the fine detail of beaches – the waves trim away the children's sandcastles and holidaymakers' footprints.

Other processes are longer term. The year-in, year-out breaking of waves moves sand and pebbles to *build* the beaches. Behind them marshes and mud flats exist, gradually being filled by sediment encouraged by the growth of plants. Lodmore and Radipole Lake shown on the map are examples

1 After reading the introduction and looking at figs. 7.1 and 7.2 list the various ways people have used this environment. Describe some of the conflicts which occur in coastal land use.

2 For an area you are familiar with (e.g. from a field visit or because your school has maps and photos) produce a sketch map similar to fig. 7.1 to illustrate how people use and have used the coast.

figure 7.2 Weymouth, an aerial view looking south west with Portland and Chesil Beach in the background

figure 7.3 Waves breaking off the western side of Portland

of this. In other places the waves erode the rocks, to produce the steep cliffs of western Portland for example (fig. 7.3).

Over a longer time period, of tens of thousands of years, not even the sea level is fixed. On Portland, raised beaches, deposits of shingle 20 m above present-day sea level, indicate that the sea was once higher than today. At the same time there are signs that the last ten thousand years have seen an overall rise in sea level. Chesil Beach is an example. This huge bar probably formed when the sea level was lower. As the sea level rose at the end of the Ice Age it was gradually driven inland. Today it links the 'Isle' of Portland to the mainland (i.e. it is a **tombolo**). Behind it is the Fleet, whose indented northern shore reflects the drowning of the river-eroded landscape of southern Dorset.

Summary

From looking at this small piece of English coastline a range of points has emerged. People use the coast as a *resource*, for transport, tourism, minerals (rock, sand and gravel). Some uses *conflict* so that the management and planning of coastal land use is an important question for society. Part of management is understanding *how the natural processes work* and realising that some of these are short term, others long term. How these processes work, how man interferes with them and is affected by them are the themes of the chapter.

Waves

The erosion, transport and deposition by the Atlantic waves was part of Surtsey's story in chapter one. Our study of coasts will therefore begin by looking at waves in more detail. They are generated by wind blowing over water. In open ocean the wave shape, its crest and trough, moves forward with the wind. Individual particles of water, however, merely move in a circular path. (This oscillation is shown in fig. 1.6A.)

How big waves become depends on three variables. Table 7.1 shows how waves of different heights (the bold figures) are produced by different combinations of wind speed, duration and fetch. The faster the winds blow the larger the wave. Large waves take some time to grow so the duration of the wind is an important factor. The third factor, the fetch, is the distance which the wind can blow over open sea. It was introduced in chapter 1 when the evolution of the northern tip of Surtsey was being described.

100

table 7.1   The three controls of wave size

II Wind speed	I Fetch		
	16 km	60 km	250 km
fresh breeze (29–38 kph)	3 hours 1m	9 hours 2m	27 hours 2.5m
gale (75–88 kph)	2 hours 2.5m	6 hours 5.5m	18 hours 10m
storm (103–117 kph)	2 hours 3m	5 hours 7m	16 hours 13m

III   9 hours = duration of wind   2m = wave height

Waves generated by winds blowing *at the time* are often called sea or *storm* waves. Waves can, however, 'freewheel' across the ocean well away from their birthplace: these are known as **swell**.

Marine erosion

The most spectacular examples of the sea's work can be found on cliffed coastlines where rocks directly face the waves (fig. 7.3). The *erosion of cliffs* can be produced by five groups of processes.

1. Firstly there is the process of wave **quarrying** or hydraulic action. The impact of waves can loosen or pull away rock. Lines of weakness, such as joints or bedding planes, are susceptible to such hammering and sucking.

2. The second process, also related to wave movement, is known as **abrasion**. This occurs when waves throw sand, pebbles and boulders against the rock, physically wearing the cliff away. The wave motion also grinds sand, pebbles and boulders together breaking them and rounding their edges.

3. The successive drying and wetting of rock (*water layer weathering*) can also weaken cliff material.

4. *Chemical erosion* by sea water may occur if the rock's minerals are liable to solutional attack, limestone for example.

5. If you have seen a rocky coastline during a storm with wind, waves, spray and the noise of pebbles chattering and chinking as they move over each other, it is difficult to imagine how important our fifth process of bioerosion can be. **Bioerosion** is the term used for the attack of organisms on the coastline. You may have noticed that many British cliffs are veneered with black lichens at the water level and orange coloured lichens above them in the spray zone. If you prowl about in the rock pools and around the tidal limits you may

1   Explain what controls wave height.

2   Consult fig. 7.1 and an atlas map which shows both shores of the Channel. For the west side of Portland and for Weymouth describe the weather conditions under which you would expect the largest and smallest waves.

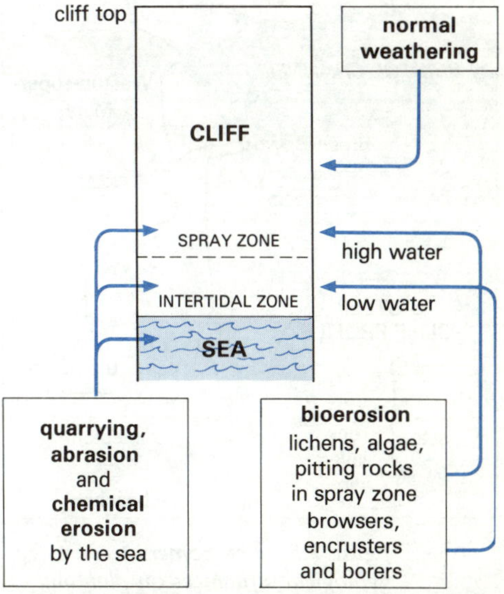

figure 7.4   Cliff erosion processes

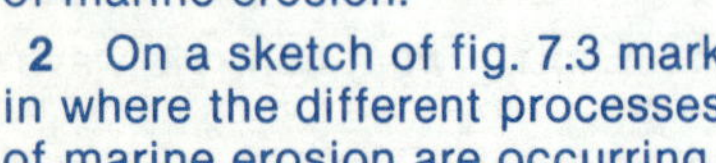

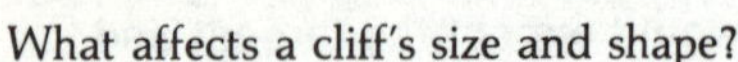

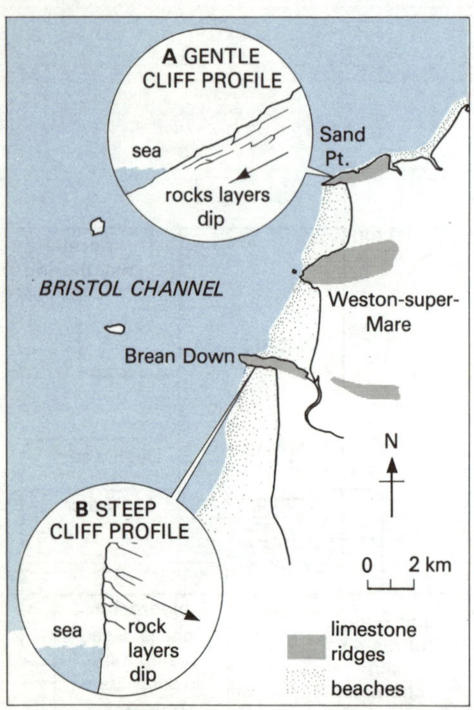

figure 7.5   The Somerset coast: geological influences on plan and profile

find many plants and animals. Some of these bore into the rock for food and shelter, others encrust its surface (like barnacles) whilst others browse over it. Humble snails, for example, scrape algae from the rock and in the process can remove fine rock debris. Season after season, generation after generation, the rock is quietly removed . . . in some locations at a rate of 1 cm in 16 years!

The erosional work of the sea is concentrated below the high water mark (fig. 7.4). Above this the rock may be exposed to the kinds of weathering and movements which were described in chapter 4.

Cliffs

Cliffs occur in an enormous range of sizes and shapes. Their *size* is influenced by the relief of the coastal area, the height of its hills and valleys. The interplay between structure (the rocks) and process (waves, weathering and mass movements) produces the variety of cliff shapes.

Where rock layers are virtually horizontal and dipping very gently inland, so that weathered blocks loosened around the bedding planes and joints do not easily fall, a steep cliff face results (fig. 7.3). Figure 7.5 shows a section of the Somerset coast and the small inserts illustrate the shape (profile) of the cliffs and the different dips of the limestones. If you look at the northern one (A) you can see that this cliff shape is gentle. The dip of the rock layers here allows gravity to pull the blocks seaward.

Geology also affects the *plan* shape of coastlines, especially where rocks of different hardnesses lie next to each other. Figure 7.5 shows this well with the three limestone headlands jutting into the Bristol Channel. The limestone ridges are more resistant than their flanking rocks. Here the 'grain' of the geological structure is almost at right angles to the overall trend of the coast.

Headlands produced by higher ridges of more resistant rock like those in fig. 7.5 can focus wave erosion on themselves. This erosion is concentrated in the intertidal zone, as fig. 7.4 showed. Waves often cut notches in cliffs at this point, especially where lines of weakness in the rock, like joints or faults, exist. If the rock above is strong enough *caves* may form. Caves eroded from either side of a thin headland may occasionally join to form arches. When the roofs of these collapse, isolated **stacks**, pillars of rock separated from the main cliff line, may form.

figure 7.6   A Dorset cliff. The rocks here are sedimentaries of different hardnesses, dipping slightly inland. Notice the falls of rock and the way this debris is becoming rounded by wave action.

figure 7.7   Lulworth Cove

figure 7.8   Land's End: a cliffed coastline cut in igneous rocks

Where rock layers run parallel to the coastline a situation like that shown in fig. 7.7 can develop. This is sometimes called an *accordant* coastline.

The last few paragraphs used examples from sedimentary rocks, but as fig. 7.8 shows, even a hard, resistant rock like granite can have its weaknesses (the jointing cracks) slowly exploited. Here we have steep, castellated cliffs, headlands and tiny pocket beaches.

1 Draw diagrams to show the formation of caves, arches and stacks.

2 Write a short essay describing how rock influences cliff and shoreline shapes.

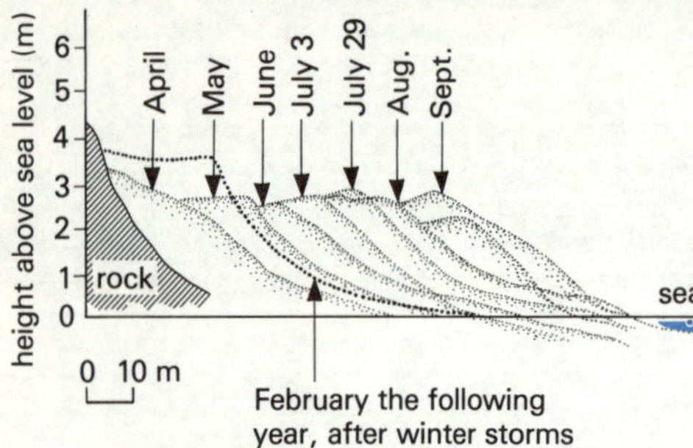

figure 7.9 *The changing shape of a beach: cross profiles of the beach at Carmel, California*

Beaches

A lay person's definition of a beach would probably be 'a strip of sand above the water'. Our British image is a beach made of yellow quartz sand grains, but in other countries the beach could be white, glistening coral fragments or black volcanic sand. The materials of the beach can cover a range of sizes (sand, shingle and cobbles) and be composed of a wide range of rock types. A beach can be formed of anything; at Fort Bragg in California Bascom mentioned one made of tin cans, washed in and built by waves carrying them from the city dump!

Fig. 7.9 shows the results of surveys on a beach in California. If you look at this you can see how the profile of the beach changes. In summer the **berm** above the high water mark is well developed. In winter this berm is smaller and a **bar** occurs below low water mark. A beach is therefore a dynamic place – its material is in constant motion. In periods of a few hours the sea can erase footprints and sand castles; over months its profile can alter, as fig. 7.9 illustrates, and over the years beach material can arrive and/or leave. The definition produced by D. W. Johnson many years ago: 'a beach is a deposit of material which is in transit either longshore or off and on shore', is therefore a good one for geographers, as it reminds us that beaches are stores of material.

With such a definition, key questions are how beach material *moves* and how waves work. In the open sea, wave shapes move forward, but not the water itself. Where the sea shallows the waves begin to 'feel bottom' and the circular motion of particles is hindered (fig. 7.10). This means that waves approaching the shore change from oscillations to

The definition of a beach

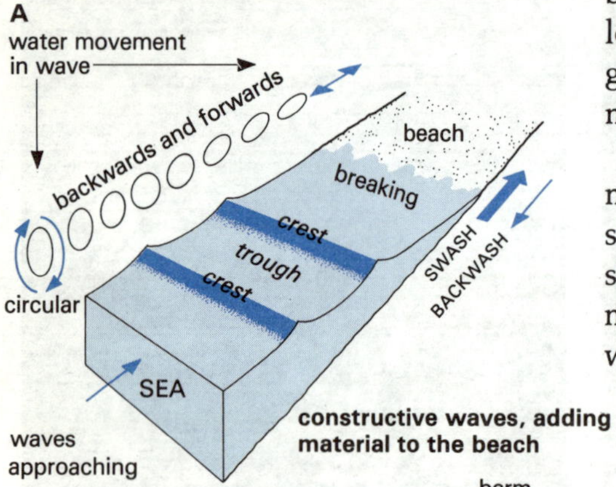

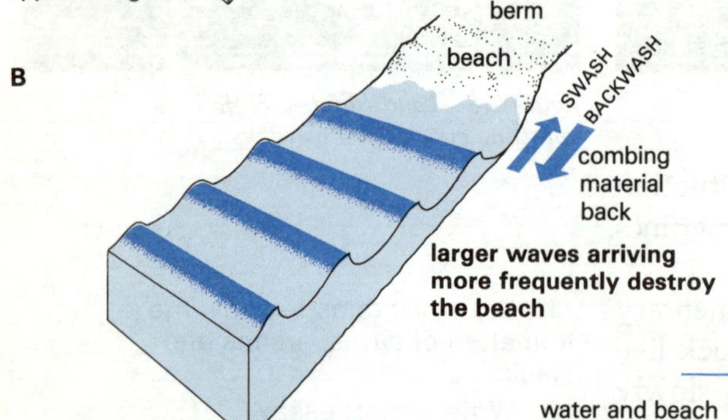

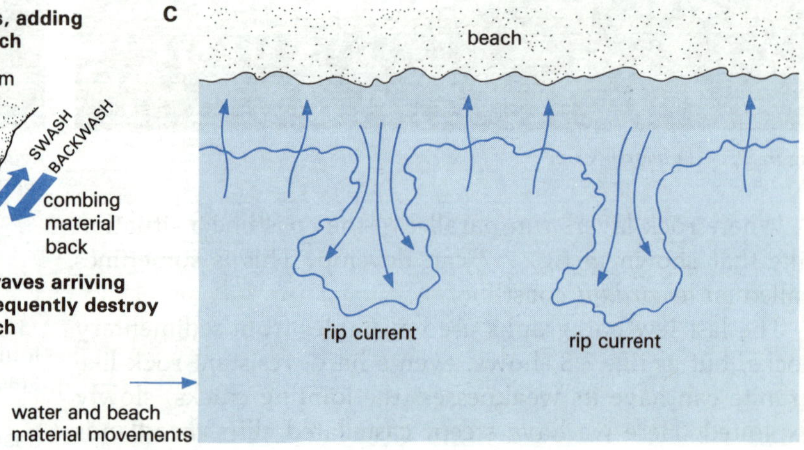

figure 7.10 *Waves and beaches*

translations of water, i.e. the water does not merely circle, it moves backwards and forwards. In shallowing water, wave crests steepen and eventually break. Water crashing down from the breaking wave's crest can dislodge and pick up beach material. Sand-sized particles are especially easy to dislodge and lift. On the beach itself water surges forwards when the wave breaks. This is known as the **swash**, the returning flow down the beach is called the **backwash**.

Waves breaking on a beach can be divided into two types (fig. 7.10). When waves arrive frequently, an incoming wave's swash is hindered by the backwash of the preceding one. The backwash is more powerful, pulling material down the beach – a *destructive* situation (fig. 7.10B). Whilst most of the foreshore is being lowered, waves may actually throw larger particles high up the beach, beyond the reach of the downwash, building a berm. Figure 7.10A shows in contrast a *constructive* situation, where waves arrive less frequently. The swash is relatively powerful, sweeping up the beach unhindered by returning backwash.

As fig. 7.9 illustrated, there may be a seasonal rhythm to wave construction or destruction. In winter heavy surf removed sand from the berm and deposited it on the bar. In summer swash carried sand up the beach. In beaches as in other things, California might be larger than life to British eyes – summer beach building extended this beach seawards at over 3 metres a day!

The description so far has been simplified. The offshore movement can actually occur as localised 'rip currents'. Such outflows of water can be dangerous for swimmers, but as fig. 7.10C suggests, they are part of the inshore-offshore exchange of beach material. Another complication is when waves approach the shore obliquely or at an angle. This situation was introduced in chapter 1 (fig. 1.6C) which described how beach material was moved sideways to build the northern tip of Surtsey. In this process of **longshore drifting** wave swash travels up the beach diagonally. (i.e. normal to the direction of the approaching wave crests) but the backwash moves back down the steepest part of the beach.

Erosion, transport and deposition on the coast

So far, cliffs and beaches have been considered separately. In nature they are parts of a whole coastal environment and cliffs, beaches and offshore areas are often interlinked. In many cases it is a self-adjusting system, able in its natural

figure 7.11   Groyne effects

state to cope with the stresses imposed by extra inputs of rock material (from occasional cliff falls, for example), large storms and so on. It is now useful to look at some examples of this interaction.

Our definition of beaches mentioned that they were deposits of rock material in transit either alongshore or off and on shore. Some beach sand therefore comes from erosion of rocks exposed as cliffs and some arrives from the streams dissecting the land. When there are prevailing winds from one direction this material may move along the coast as a virtual 'river of sand'.

Such rivers of sand may be interrupted by natural barriers like headlands, indentations like river mouths and by human intervention. You may have seen signs of this interruption of the beach conveyor belt on some English beaches where **groynes** have been built. These are barriers of wood or steel placed fence-like to run down the beach. The beach drifting process means that sand and shingle pile up on one side of the groyne. A series of groynes along a beach can hold back the longshore movement and increase the thickness and width of a beach. This may make it a better holiday beach and it may also protect the backshore from erosion. Where a beach is backed by easily eroded cliffs, a thicker and wider beach protects them from undercutting and collapse. However, further along the coast beaches may be starved of fresh beach material to replace that still being lost by longshore drifting. Beaches may thin and expose the cliffs backing them to increased erosion.

Wave action and longshore drifting often sort out beach material and deposit them in different positions according to their size. This happens up and down a beach; only powerful storm waves can move larger pebbles to build a storm ridge at the top of a beach, for example. The same sorting occurs *along* beaches too. Chesil Beach, which was mentioned in the introduction, is an example. At the Portland end, the mean size of pebble is 5.9 cm, at point X on fig. 7.1 it is 3.36 cm and at the western end of the beach, at West Bay (Bridport) the material is pea-sized.

This natural sieving by the waves makes beaches an attractive source of aggregates, the term used for sands and gravels. Chesil is exploited for this purpose on a small scale. The pea-sized gravel at Bridport, for example, is used for road surfacing, but the rounded particles when wet are very slippery. So slippery that some were exported to Italy for the Pirelli company – design a tyre to cope with this surface and you would have a good tyre!

Removing beach material like this can upset the coastal system. A dramatic example occurred in Start Bay, Devon. Between 1897 and 1902 shingle was removed from the foreshore for building Devonport. The beach was lowered by about 6 m. The beach had protected a platform of highly fractured rock, mica schist, on which the small fishing village of Hallsands had been built. In January 1917 a combination of factors occurred. There were high spring tides and, unusually for the Channel, a north-easterly gale. Forty-foot waves battered Hallsands, causing its abandonment. The system here in Start Bay was a closed one, the shingle not being renewed from present-day cliff erosion.

figure 7.12 Barton-on-Sea: the failure of a cliff protection scheme. The central part of Christchurch Bay suffers from high erosion because of wave patterns and the soft sands and clays forming the 34 m high cliff. Engineers had put a barrier of interlocking sheet piles with drains to stabilise the upper cliff, and rock trenches with revetments to protect the cliff foot. Major slips and erosion occurred in 1974–75, destroying the expensive scheme and some cliff top property.

1 Modern society needs sand and gravel. River floodplains and the coast are two sources. List the advantages and disadvantages of these two sources.

figure 7.13 Hallsands. The abandoned village is a consequence of man-induced erosion.

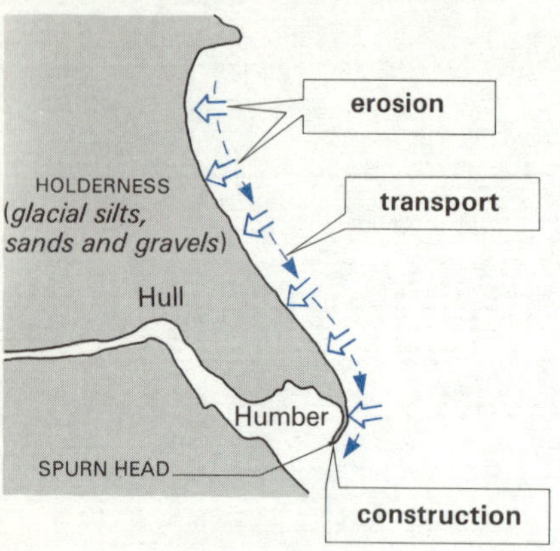

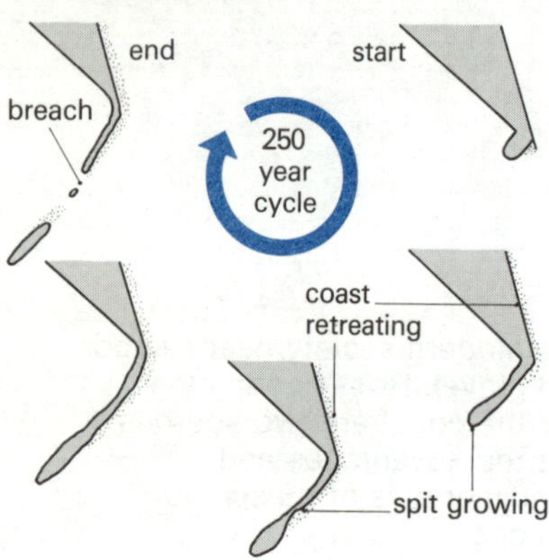

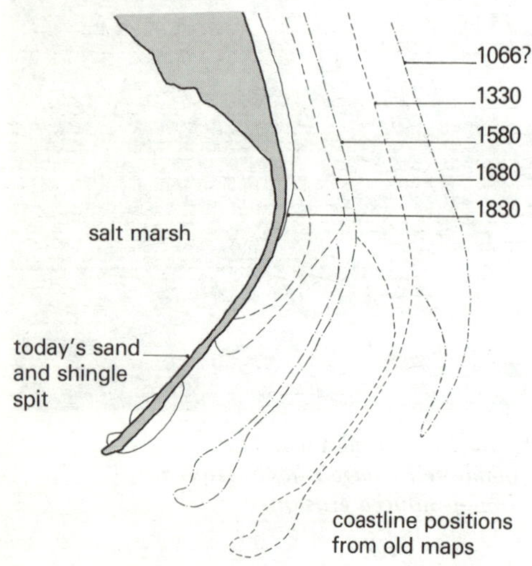

figure 7.14 *Spurn Head and the Holderness coast*

The longshore movement of material can build landforms. Spurn Head, a sand and shingle **spit** curving 5½ km across the mouth of the Humber, is an example. Figure 7.14 shows how the buildng of this spit is linked to the 60 km shoreline to the north. This Holderness coast is being eroded at one of the highest rates in the world. Annual rates of up to 2.75 m a year have been found, but with considerable variation . . . a day's retreat of 6 m was recorded in one storm.

A reason for this rapid retreat is that the North Sea is able to erode easily the soft glacial deposits of silt, sand and gravel. This softness, however, is only part of the story. If you think of it, when a cliff collapses the debris ought to protect its foot from further wave attack, at least for a while. In Holderness, however, the debris from the cliff fall does not provide much protection for two reasons. Firstly, only 10 per cent of the material is sand-sized and available to nourish the beach. The finer material is carried away quickly by the sea in suspension. For every 10 tonnes of sand moved, a supply of 100 tonnes of cliff silt and sand is needed to replace it. Secondly, the beach sand itself is also being moved southwards. If you think back to our discussion of wave sizes and you look at the shape of the North Sea you can appreciate why there is a dominant southward movement.

figure 7.15 *Westward Ho, Devon. Behind the spit and its dunes 'bioconstruction' proceeds*

'Canute' takes on the waves

By Michael Parkin

SPURN Head, a comma of land punctuating the Humber estuary, is to be closed to traffic from the middle of next week while an attempt is made to save it from the sea.

The three-mile long peninsula has disappeared four times in the past 1,000 years. What tended to happen was that the sea eroded the northern end, and then broke through.

The rest of the headland dwindled into an island and finally disappeared. Eventually Spurn Head re-formed itself from deposited material a little to the west, and there it stayed until the sea once again destroyed it.

Major Ian Kibble, executive officer of the Yorkshire Naturalists Trust, said that his organisation and the other occupants — the Coastguard, the Spurn Lifeboat, the Humber Pirates and Trinity House — were combining to do what he called "a King Canute" on the head. The Trust has a bird reserve on the headland.

The point most threatened is a narrow neck of land, about a mile down the peninsula, where the road rests on boulder clay. Tank-traps laid in the second world war, consisting of 5ft cubes of concrete, were being moved to form a sea defence.

"They will be laid in a higgledy-piggledy fashion to present a great variety of surfaces to the waves," he said. "In this way they should absorb much of the wave energy."

In the long run, though, the sea is expected to win. It very nearly broke through at the narrowest point during high tides about two weeks ago. One man on the head reckoned that, had the wind been force five or more, there would have been a risk of a breakthrough.

While work goes on, visitors will be allowed in on foot to see the bird reserve, a staging post for thousands of migratory birds flying to and from Britain.

Spurn Head to the south has been built by waves moving sand southwards across the mouth of the Humber. It is a sand and shingle spit. Its higher parts consist of sand dunes colonised by marram grass and sea buckthorn. On the sheltered western side is a salt marsh. George de Boer has studied the development of Spurn Head from old charts and the history of the lighthouses on its tip. As it grows from a receding shoreline it is forced into a 250-year cycle of growth and destruction. Figure 7.14B shows a **model** of this cycle.

Spits built by longshore drifting are not uncommon shoreline features. Figure 7.15 is an example from the Devon coast. The main body of the hooked spit can be seen on the photograph together with the sand dunes along its crest. In the sheltered water behind it deposition is occurring. Just as organisms eroded cliffs (page 101), plants may also encourage deposition. One agent of this **bioconstruction** is spartina grass which invades the salt marshes. It slows water so that sedimentation is encouraged and its roots anchor the fine muds and silts. Gradually less demanding plants invade and man too lends a hand by embanking such salt marshes and incorporating them into land itself.

As we have seen from Spurn Head and Westward Ho where the coastline changes direction (i.e. with a river mouth or estuary), and where there is a supply of longshore drifted material, spits may develop. Sometimes the sand and shingle ridges may be built to effectively join an island to the shore. This feature is known as a **tombolo**.

People often have a 'Canute' like attitude to coastal protection. If your home or community's roads are being consumed by the sea, it's understandable to cry 'help' to civil engineers. Now that we know more about coastal erosion, a longer term approach should be the *control* of land use.

1 Describe a spit and the processes which have built it.

2 Read the article by Michael Parkin. Explain why 'in the long run the sea is expected to win'.

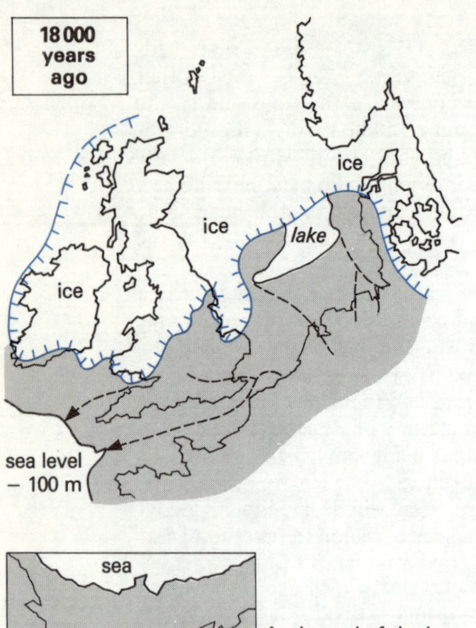

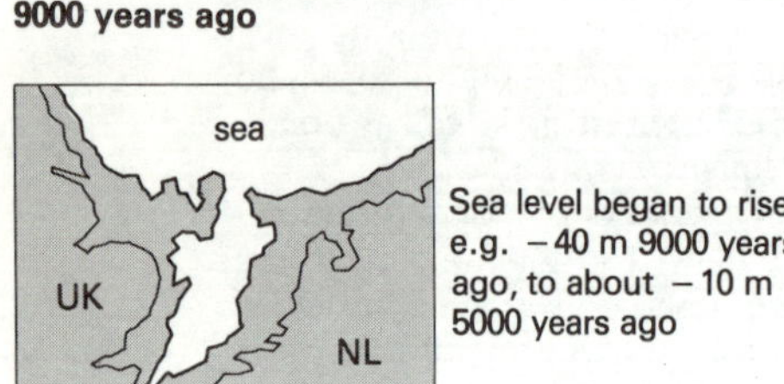

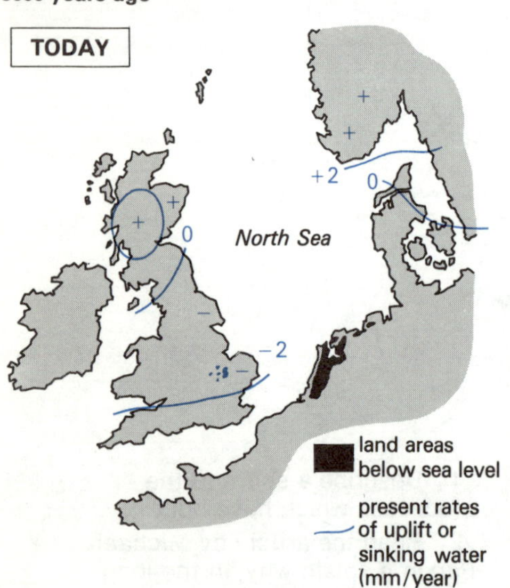

figure 7.16   The changing North Sea

Ups and downs: shorelines and sea level changes

From a study of Ordnance Survey Maps you will have seen that contours and spot heights in Britain are related to 'sea level' (based on observations at Newlyn in Cornwall). However, if we extend our view to spans of geological time, 'sea level' has hardly been a fixed datum as the relative heights of land and sea have changed. This can be seen dramatically in the changing outline of the North Sea, fig. 7.16.

Thinking back to chapter 6, you have probably already thought of a reason for this sea level rise: one connected with the decay of the Pleistocene ice sheets. At the maximum of the most recent glaciation, about 20 000 years ago, the continental ice sheets and glaciers stored so much of the world's water that sea level was about 200 m lower than today. If you look at atlas maps which show sea depths you can find areas of shallow seas today (like the Bering Strait) which would have been land at that time and islands which would have been connected together (like Indonesia). When all the water locked up' on land as glaciers and ice sheets began to melt there was a **eustatic** rise in sea level around the world.

The shapes of the coastlines produced by such eustatic 'drowning' depended on the terrain, or relief, of the land areas being inundated by the rising sea. In *lowland* areas, broad shallow estuaries with innumerable branching tributaries formed as the sea flooded the valleys and lower ground. An example of such a drowned lowland coastline can be seen on the mainland coast of the USA inland from Pamlico Sound. The shores of the North Sea (fig. 7.16) also have examples of such drowning and we saw a small-scale example in the Fleet (fig. 7.1). As the eustatic rise in sea level slowed, the shapes of such shallow inlets could become modified by the 'bioconstruction' processes mentioned earlier with the growth of mud flats and salt marshes. Man too may alter the shapes of such coasts. The 'man made' nature of much of Holland can be seen from fig. 7.16.

Where such a sea level rise occurred in more dissected hilly terrain, in plateaus cut into by river systems for example, the lower parts of the valleys became drowned. Figure 7.17 shows the resulting landform – a **ria**. Their shapes are similar to the drowned valleys of a lowland coast, but rias are deeper, with steeper sides. They can provide good natural harbours. However, being found in locations like West Ireland, Brittany and Devon and Cornwall they tend to be distant from industrial areas.

Where the 'grain' of the relief, the trend of the hill tops and valleys, is parallel to the coast, a sea level rise can produce what is often called a *Dalmation* coastline. Dalmatia, Yugoslavia's Adriatic coastline, provides the classic example, with long narrow islands paralleling the coast. (Figure 8.11 also shows an example from California. Here a breach in the coastal ranges, the Golden Gate, separates the elongated San Francisco Bay from the sea.)

The scenery of many uplands in the middle and high latitudes has been modified by ice. Valley glaciers carved the straightened and trough-like valleys described in chapter 6. When such glaciated uplands form a coastal area, rising sea levels submerge the lower parts of the troughs producing a **fjord** coastline. Figure 7.18 is a Landsat satellite image of part of the Norwegian coast. It is a May scene, so the mountain plateaus, some over 2000 m high, are still covered with snow. (This appears white in the scene. The lower snowfree ground appears as grey tones and the sea as black.) Unlike rias, fjords are relatively straight-sided, and as you can see from the image they do not narrow inland. Some of those in the image are 20 km wide even 80 km from the Atlantic. The closeness of the snow-covered high ground to the water in the fjord shows how steep their walls can be. Many fjords are often deeper in their central and inner parts than at their mouths, which reflects the ability of the valley glaciers to carve basins into their bed, just as they do to produce ribbon lakes. The shallowing at the mouth is known as the *threshold*. Fjorded coastlines similar to Norway's are found in British Columbia,

figure 7.17   A ria, Milford Haven

1   Why has sea level changed?

2   Define and describe the following: ria; estuary; dalmation coast; fjord; raised beach.

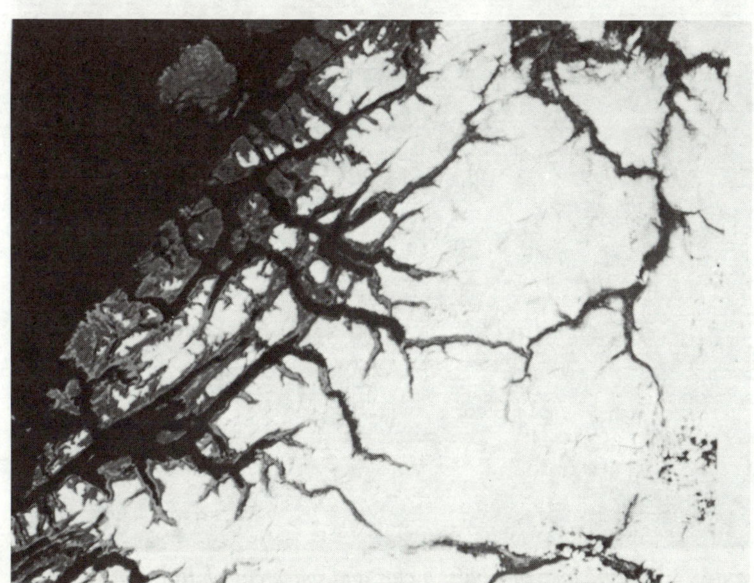
figure 7.18   A Landsat image of the Norwegian Coast

Alaska, South Island New Zealand and in a subdued form in north-west Scotland, where drowned glacial troughs are known as sea lochs.

So far the examples have all been of rising sea levels. There are, however, areas where the land has risen relative to the sea. If you look at fig. 7.16 you can see there are some areas marked by plus signs; these are slowly rising at the present time. Land can rise for a number of reasons; the earth's crust could be responding to tectonic forces, for example. The uplift of Scotland and Scandinavia, however, is another legacy of the ice sheets. Ice might be less dense than rock but the mass of an ice sheet does affect the crust beneath; put bluntly, it sags under the weight! When the ice melts the crust is unloaded and gradually 'rebounds'. This movement is called glacial **isostatic** recovery. Geomorphologists have discovered that parts of North Canada rose at rates of up to 10 m in a hundred years! Today North Sweden is rising at a rate of 900 mm and the Scottish Highlands at 300 mm a century.

Where uplift is greater than the eustatic sea level rise, **emerged shorelines** result. Uplift means that a range of coastal features, such as cliffs, caves, arches, stacks and wave-cut platforms, representing the erosional work of the sea, and beaches are found above existing shores (fig. 7.19). As time passes these become weathered and eroded and lose their freshness.

figure 7.19   Evidence of uplift: a cliff and wave-cut platform at Little Cumbrae, Bute

112

Summary

In the chapter we have looked at the following:
- The coastal environment provides a variety of resources.
- The sea is an agent of erosion, transport and deposition.
- The detail of the coast, its shapes, plan and profiles, reflect the interaction of the sea and rocks; often with mass movement, weathering, rivers, glaciers contributing to the present appearance of the landforms.

Exercises

Examine fig. 7.20 which gives information about the River Cuckmere on the Sussex coast.

14000 YEARS AGO (sea level about 100 m below present)	Tundra conditions. Cuckmere a periglacial river (see chapter 6) with a braided channel cutting a 750 m wide valley. River a tributary of the 'Greater Seine' which drains westwards from the English Channel.
8000 YEARS AGO (sea level about 20 m below present)	Valley flooding begins. Ria forms. Clay soils of river's basin protected by tree cover flourishing as climate warms. Man begins clearing forests
2000 YEARS AGO (sea level close to today's)	Forest clearance in basin, for timber and plough, increasing, leading to increasing erosion of soils and deposition in the ria. Ria silting continues, narrows into a meandering estuary, with tidal mud flats. Spit grows across mouth, which increases sedimentation behind it. River keeps breaking through spit (see map).

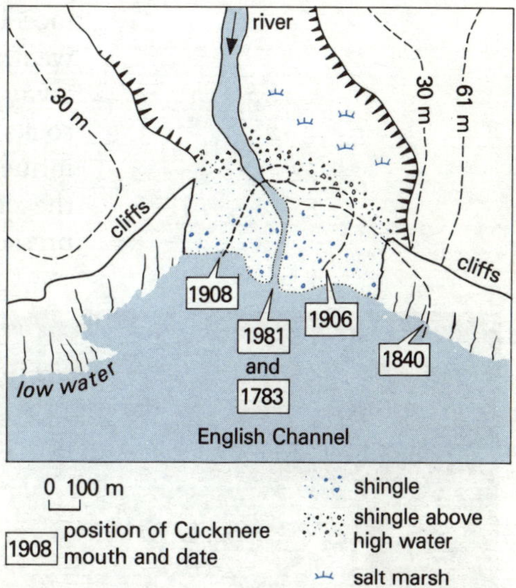

figure 7.20   *Changes at the mouth of the Cuckmere River, Sussex*

1   Draw a series of diagrams to illustrate the changes over the last 14000 years.

2   The 1981 mouth of the river is artificial. Describe the changing positions of the river's mouth and suggest why you think these may have happened.

Figure 7.21 is a model of a coastal system.

3   Give a brief description of all its inputs and outputs.

4   List the ways in which man has affected these inputs and outputs or been affected by them.

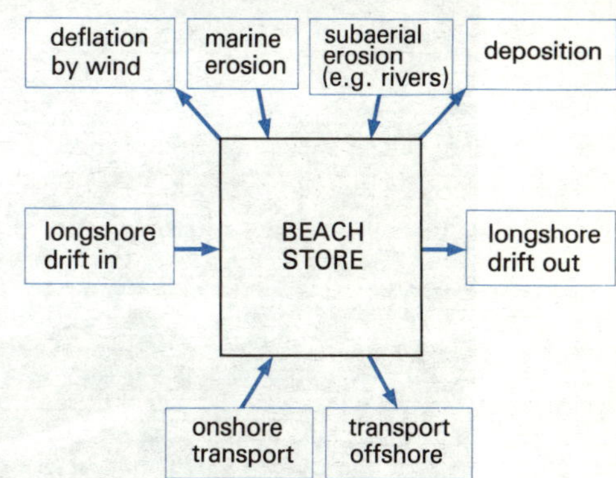

figure 7.21   *Beach systems*

8 Rocks and scenery

Introduction

Imagine you are an astronaut looking down on the area in fig. 8.1. Few details of human activities are visible. What stands out is the erosion of crumpled rocks. The earth's surface has been 'out of doors' a long time so the agents of weathering, erosion and deposition have been able to carve away, move and dump the materials of the earth's crust. Its rocks have different compositions and hardnesses, which influences landforms. The other geological influence is that the crust is dynamic, with up, down, stretching and compressing movements.

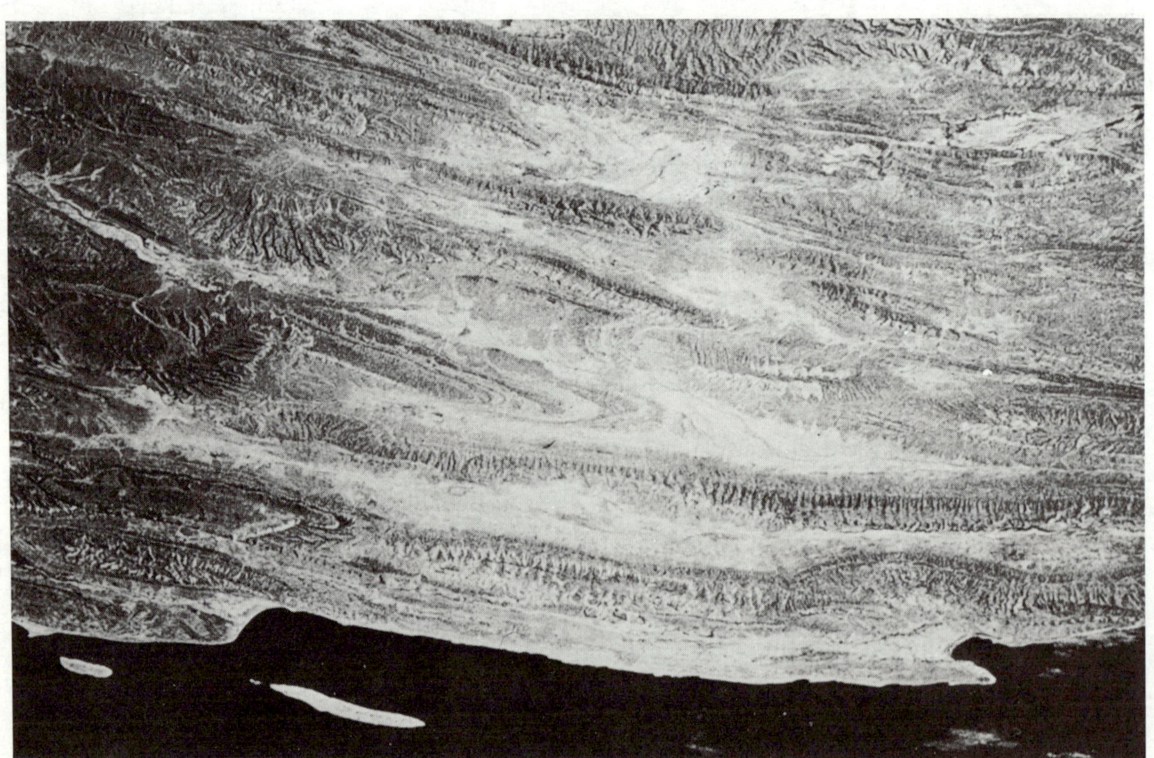

figure 8.1   *A satellite view of folded sedimentary rocks which have suffered erosion: the Zagros Mountains, Iran*

114

The erosion of folded sedimentary rocks

The rock cycle was introduced in chapter 1 (fig. 1.11). Part of the cycle is the deposition of material to form sedimentary rocks. Figure 8.2A shows a sequence of sedimentary rocks from the Welsh borderland. The oldest rocks in this geological column (the Llandovery Sandstones, about 400 million years old) are found at the base. Above them is an alternating sequence of limestone and shales. All were deposited in layers (**strata**) across the seabed. Our story however does not end here, as these rocks were compressed and later bent, or arched, into a dome.

Figure 8.2B shows the landscape today, after 60 000 000 years of weathering and erosion. The outermost ring of hills is composed of the Aymestry Limestone, which originally formed a **dome**, with rocks sloping away in all directions. In the beginning streams would have flowed radially outwards from its centre. Gradually they cut their channels into the dome and their valleys widened. In this way the underlying rocks would have been progressively exposed. These responded differently to erosion. The shales are relatively soft and they generate more surface water flow than the adjacent limestones. Therefore the shales now underly the vales, where tributary streams were able to extend their valley lines. The limestones remain to form the higher ground (fig. 8.2).

Relative hardness, i.e. resistance to streams and slope processes, is only part of the story. The tilt of the rocks (the **dip**) is equally significant. Where the limestones dip gently, asymmetrical ridges develop: the **scarp** landform with its gentle dip slope and steeper scarp slope. Where the strata dip more steeply (as on the west side of fig. 8.2) an almost symmetrical steep-sided ridge or **hogsback** results.

The landscape shown in fig. 8.2B therefore reflects a combination of geological factors:

1. the relative hardness and softness of the different rocks;
2. the variation in the dip of the layers, or strata;
3. the overall shape of the structure.

The original dome has been transformed. Its successive layers have been peeled away. The limestones, doing this more slowly, stand up as curving ridges interleaved with the shale-floored vales. The oldest rocks outcrop today at the centre of the dome where the relatively resistant sandstone and limestone stand up as a raised 'boss'. The Woolhope Dome, although only 7 km across, therefore shows the influence of rock hardness and geological structure on scenery.

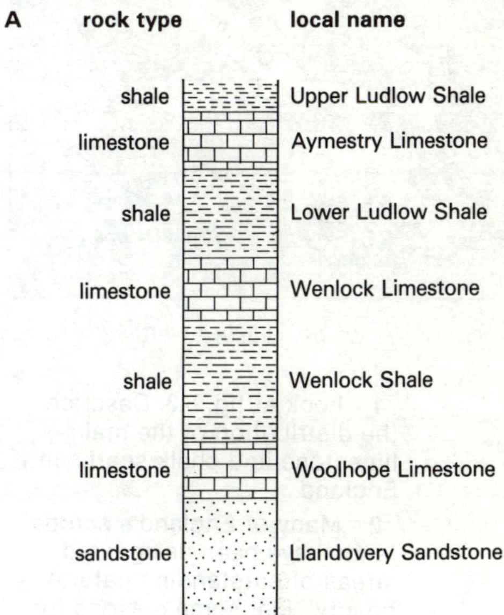

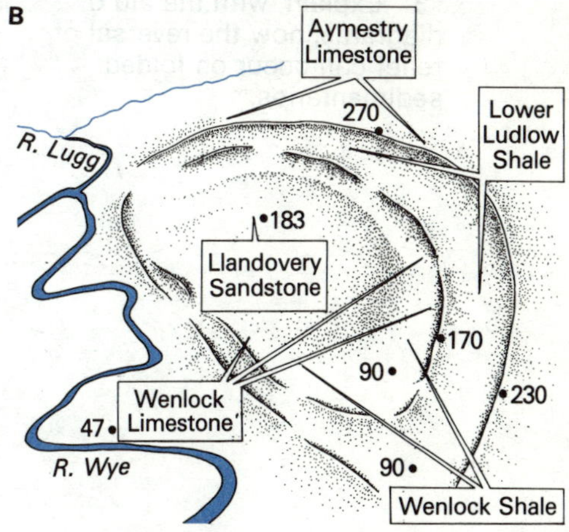

figure 8.2 *Sedimentary rocks in the Woolhope Dome, Hereford*

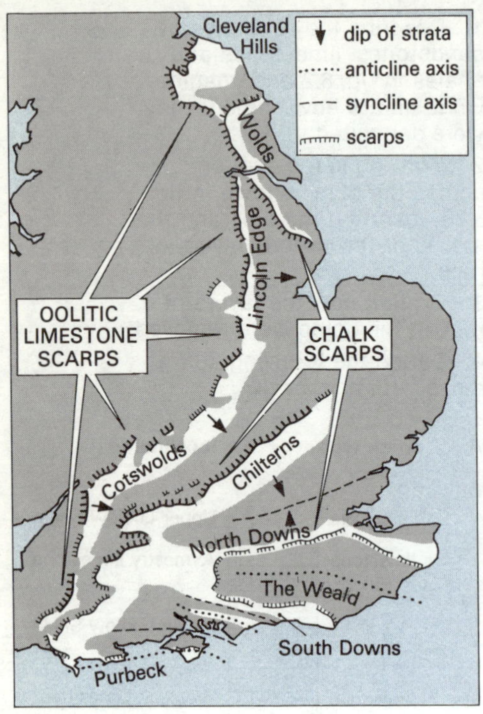

figure 8.3   English scarplands

figure 8.4   Limestone scarps in Shropshire

1   Look at fig. 8.3. Describe the distribution of the main limestone and chalk scarps in England.

2   Many of England's scarplands have been designated 'areas of outstanding natural beauty'. List some reasons for this.

3   Explain, with the aid of diagrams, how the reversal of relief can occur on folded sedimentaries.

Scarps occur wherever gently dipping strata of different resistances are adjacent (fig. 8.2). Figure 8.3 maps the distribution of the main chalk and limestone scarps in

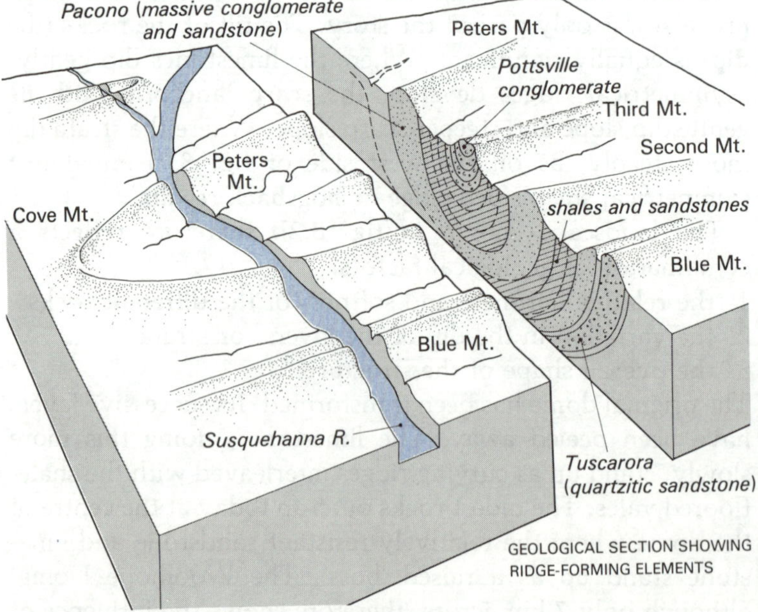

figure 8.5   Block diagram and geological section of the Appalachian Mountains

England. It also shows the Weald in south-east England. This large arch (or **anticline**) of chalk has been eroded in the same way as the Woolhope Dome.

In domes the stretching of rocks in the arch can weaken them. In the trough (or **syncline**), on the other hand, strata are compressed. When erosion and rock removal is prolonged the situation shown in fig. 8.5 can result. If you look at the section line you can see that Third Mountain is the remnant of a syncline. The relief is 'reversed', the original syncline trough now forms a *ridge* of higher ground. This figure also shows the tapering folds: it's easy to forget the three-dimensional form of geological structures!

Figure 8.6 shows a more complicated situation. Here compression has been so great that the strata have fractured and slid over and underneath each other along 'fault planes'.

Folding and mountain building

Two facts about intense folding are intriguing. Firstly, in fig. 8.6 for example, limestones form many of the mountains. Today's descendants of the organisms which formed those limestones can live only in the sunlit upper few metres of the sea – yet limestone thicknesses many hundreds of times that are found! This paradox can be explained by imagining a sea floor which *sinks* as the sediments accumulate. In the case of the Appalachians (fig. 8.5) geologists have recognised 13 000 metres of such shallow water sediments! Such a basin with its gradually deepening floor is known as a **geosyncline**. known as **creep**. It was creep which was damaging the winery example, the marine deposits are now found 4000 m above sea level. In many cases mountain peaks are parts of synclines or downfolds (fig. 8.5 and 8.6)!

These facts suggest that the process of mountain building has two parts. Firstly, the creation of a deepening basin, or geosyncline, within which sediments accumulate. These sediments are then compressed and contorted when the geosyncline closes. (This happens as the earth's crustal plates move.)

The tops of the crumpled folds are attacked by agents of weathering and erosion. Figures 8.1 and 8.2 showed ample signs of this. Such erosion, by rivers, ice and waves, can be so effective that only the 'stumps' of folds remain. The removal of rock material on such a massive scale upsets the balance of the crust so that vertical *uplift* begins. The removal of rock has the same effect as the removal of the weight of an ice sheet. The process of **orogenesis** (or mountain building)

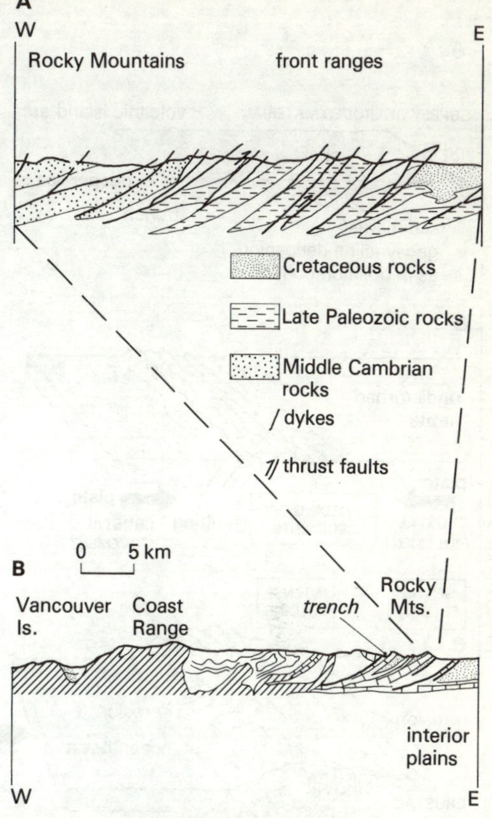

figure 8.6   *Folding and faulting in the Rocky Mountains, western Canada*

therefore reflects the operation of *two* forces – a compressional one and an uplifting one.

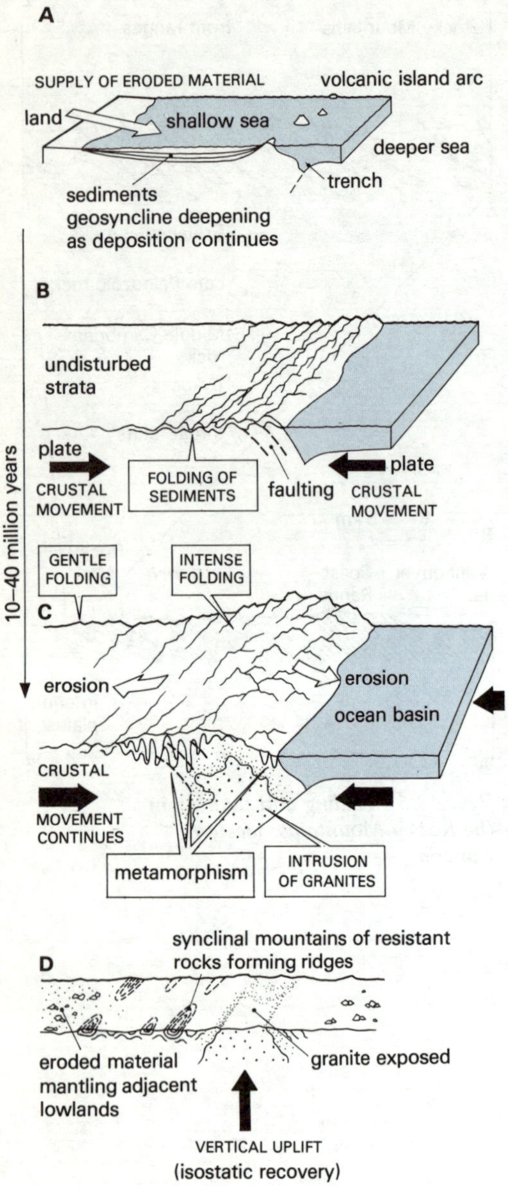

Observations from mines and boreholes have shown that the temperature of rock increases with depth, at a rate of about 3°C/100 m. This means that sediments forming the 'roots' of the fold systems (i.e. at depths of 60 km) are subjected to very high temperatures. The pressure is also much higher than on the surface. Under such conditions the sediments become *metamorphosed* or altered (see page 47). The melting of crustal sediments can produce volcanoes and magma may also be injected between the fault planes to produce features like the gabbro **dykes** shown in fig. 8.6. Magma may also cool slowly at depth and form granite masses. These **batholiths** may be exposed at the surface as the overlying rocks are stripped away.

The process of orogenesis is summarised in fig. 8.7. Orogenesis has occurred at distinct periods of the earth's history and it has affected different parts of its surface. Figure 8.8 includes both dimensions. The time scale shows the main mountain building episodes, together with the names which geologists have given them, and the map shows their distribution. The most recent episode, the Alpine Orogeny, is the freshest as the processes of weathering and erosion have had less time to remove the effects of the compressional and uplifting forces. The exposed parts of the oldest systems are now relatively stable areas of old hard rock, or **shields**.

1 Describe the processes of fold mountain building summarised in fig. 8.7.

2 Examine fig. 8.8 and describe the distribution of the recent fold mountains.

3 Mountains are barriers to movement. They receive more rain and snow as air is forced to rise (and cool) over them. This makes them a resource for water and winter sports for example. They also act as barriers between peoples. Take one of the fold mountain areas shown in fig. 8.8 and, using atlases and other reference books, produce a series of sketch maps to show these two effects.

Faulting and earthquakes

Folding is not the only type of earth movement. Figure 8.9 shows evidence of another kind. The Cienega Winery is being gradually torn in two!

The horizontal movement is the obvious one. There has been, however, a small amount of vertical movement, as you can see from the miniature scarp. This area is part of the impressive San Andreas **fault** (fig. 8.10) which is the most

figure 8.7 Mountain building

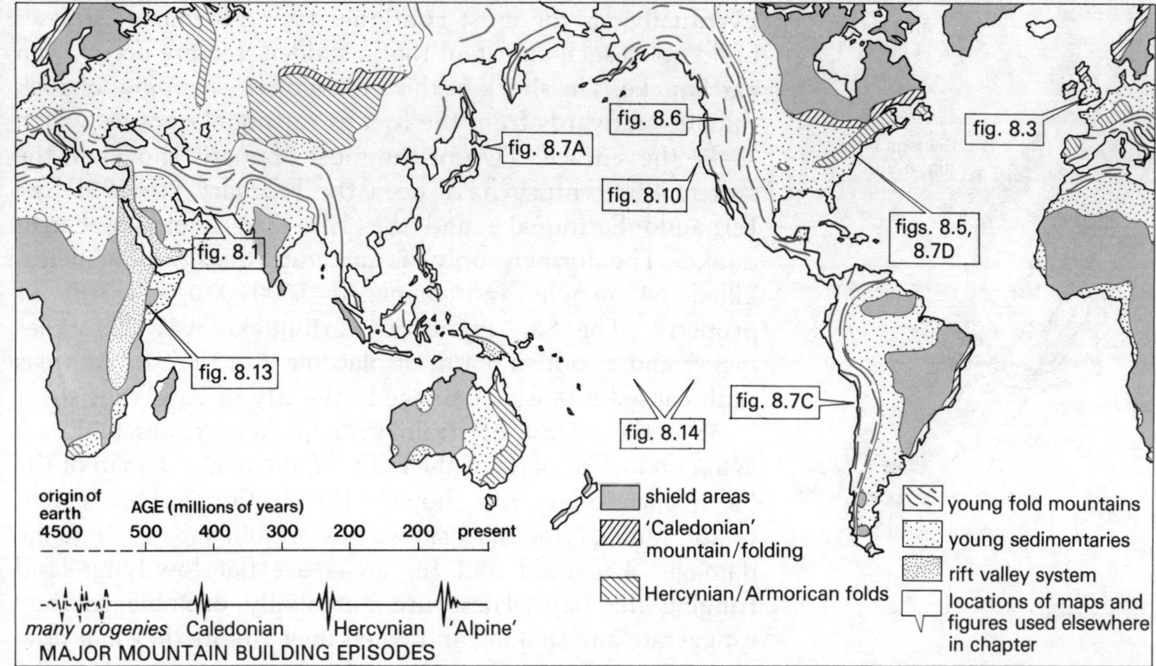

figure 8.8   World fold mountains

closely watched fault system in the world – not surprising when you see that it passes beneath the most densely populated parts of California. Figure 8.10 also shows that the earth's crust to the west, the Pacific Plate, is moving north-north-westwards. The present speed of this movement is about 60 mm a year. When this movement occurs with frequent small slippages and fracturings of rock the process is known as **creep**. It was creep which was damaging the winery in fig. 8.9.

However, when the rocks on either side of the fault bind together because of friction the fault becomes locked.

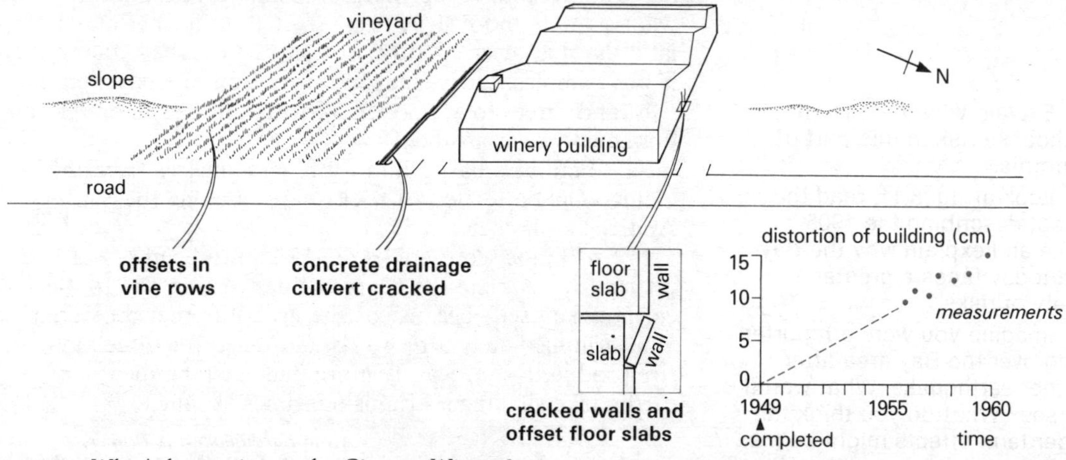

figure 8.9   What's happening at the Cienega Winery?

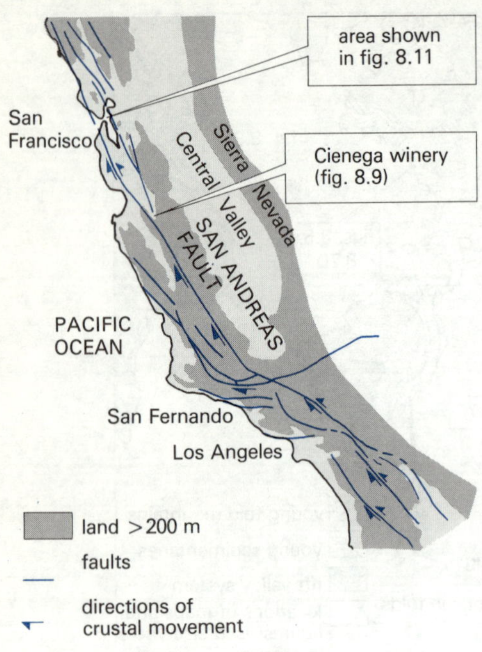

figure 8.10   The San Andreas fault system, California

Eventually, as the crust continues trying to move, stresses build up to such an extent that a sudden fracture occurs – an **earthquake**. The shock of this breaking, or snapping, of rock radiates outwards from the foci as earthquake waves which shake the surface. Two of the most dramatic shocks in the system this century have been the February 9th 1971 San Fernando Earthquake and the 1906 San Francisco Earthquake. The former, only a moderate quake, nonetheless killed 64 people and damaged $500 000 000 worth of property. The San Francisco Earthquake was 350 times bigger and a sudden 6.4 m displacement of the San Andreas fault caused extensive damage to the city of San Franciso.

After the release of strain in a quake, stresses build up again on locked parts of the fault. Figure 8.11 is a map of the San Francisco Bay area showing the position and movement of the faults. This map shows susceptibility to earthquake damage. The mud and fill areas are flat low-lying land fringing the bay. These are potentially unstable as they exaggerate any shaking and when they vibrate they can flow almost like a liquid. The other hazardous zones are where the bedrock is liable to *slide* when shaken. (Slides and flows were discussed in chapter 4.) Unshaded are the bedrock areas best able to withstand shaking during an earthquake.

The 1906 San Francisco Earthquake

April 18th, 1906, early morning in California. By the Golden Gate slumbered San Francisco, a city of 400 000 people. Built in a series of economic booms during the previous century, it was a mixture of old and new buildings, all constructed with little heed to natural hazards. Already the downtown area was dotted with steel-framed high rises, but it was still dominated by older buildings of wood and unreinforced brick that lined the narrow streets and unprotected openings. Around the wharves were more structures, erected on former marsh land that had been used for so long as a garbage dump that it was completely dry. Further away from the bay were the two- and three-story wooden Victorian homes, more elegant but equally combustible.

At 12 minutes past 5 a.m., a few kilometres from the Golden Gate a section of rock snapped along the San Andreas fault.

In San Francisco there were 315 known deaths. But it is difficult to estimate the degree of destruction due to the earthquake itself, because of the great fire that broke out almost immediately afterwards and raged for three days ... fire produced significantly more damage than did earthquake shaking, perhaps ten times as much.

from *Earthquakes: A Primer*,
Bruce Bolt (W.H. Freeman, 1978)

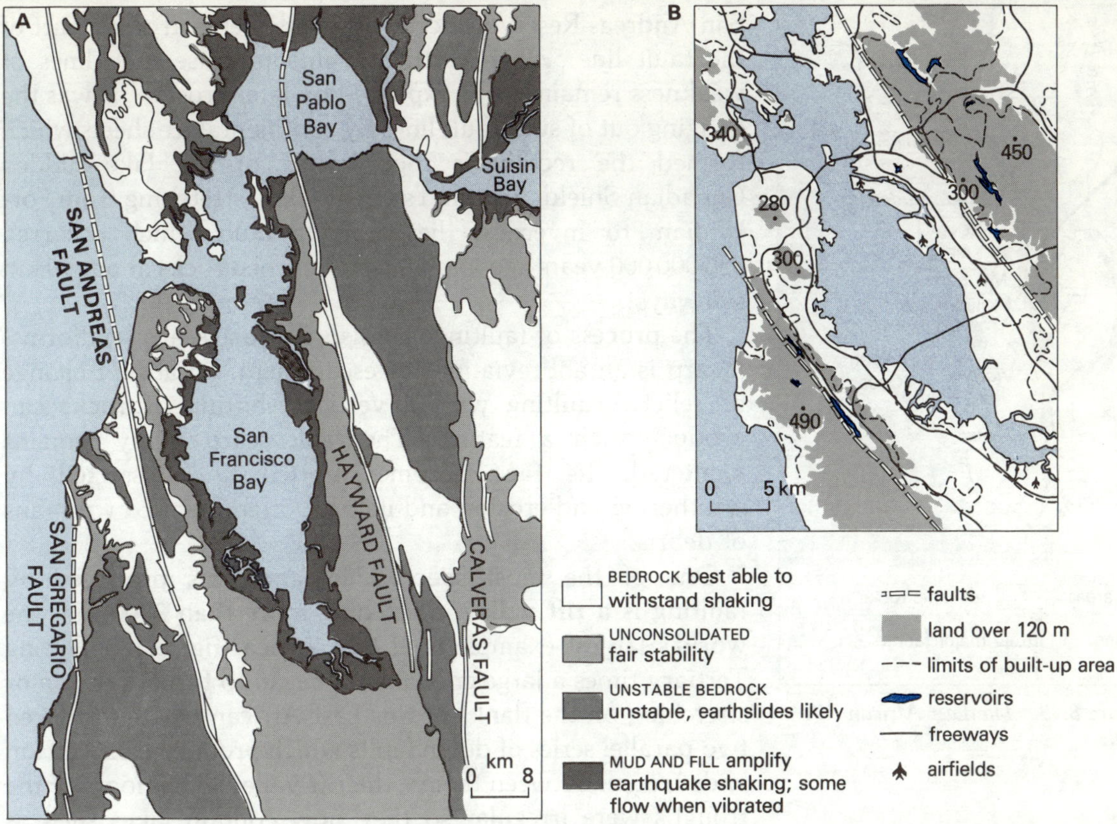
A
San Pablo Bay
Suisin Bay
SAN ANDREAS FAULT
San Francisco Bay
HAYWARD FAULT
CALVERAS FAULT
SAN GREGARIO FAULT
0 km 8
BEDROCK best able to withstand shaking
UNCONSOLIDATED fair stability
UNSTABLE BEDROCK unstable, earthslides likely
MUD AND FILL amplify earthquake shaking; some flow when vibrated
B
340
280
300
450
300
490
0 5 km
faults
land over 120 m
limits of built-up area
reservoirs
freeways
airfields

figure 8.11 The geological hazards of the San Francisco Bay area

The kind of damage likely with an earthquake as powerful as the 1906 shock has been estimated by the US Office of Emergency Preparedness. Buildings like the Transamerica Tower (fig. 8.12) have been designed to withstand shaking and not shed their windows and masonry down to the street. The various bridges and the BART subway tube under the bay are also designed to withstand earthquakes. Nonetheless the next major movement of the San Andreas fault, in an area containing 4 700 000 people, will produce casualties. Their number will vary according to the time of day. A quake at 2.30 a.m. would probably kill 3000 and hospitalise 11 000; a quake at 4.30 p.m. might kill 10 000 and hospitalise another 40 000. If the nine reservoir dams in the area fail, the dead and injured toll could rise to over 100 000 people.

Important as earthquakes are to people living in active fault zones, it is their significance in shaping the landscape which is our theme here. At the fault plane where the crust is moving, rock may be split and fractured into a mixture known as 'gouge'. This forms a line of weakness which the agents of weathering and erosion may exploit. In fig. 8.11 the

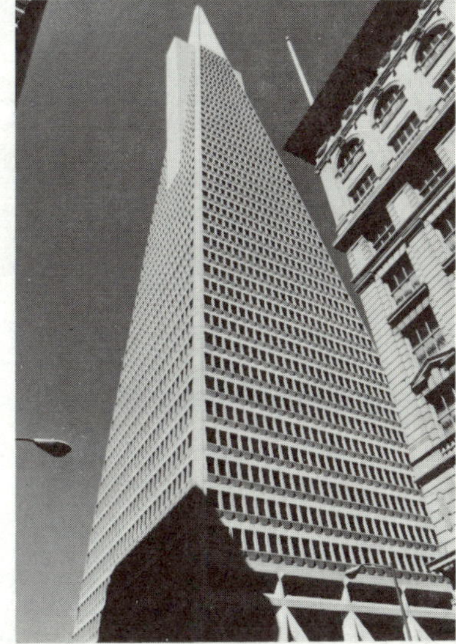

figure 8.12 San Francisco: earthquake risk and office block design. This building is designed to resist shaking and not shed any window glass onto the street.

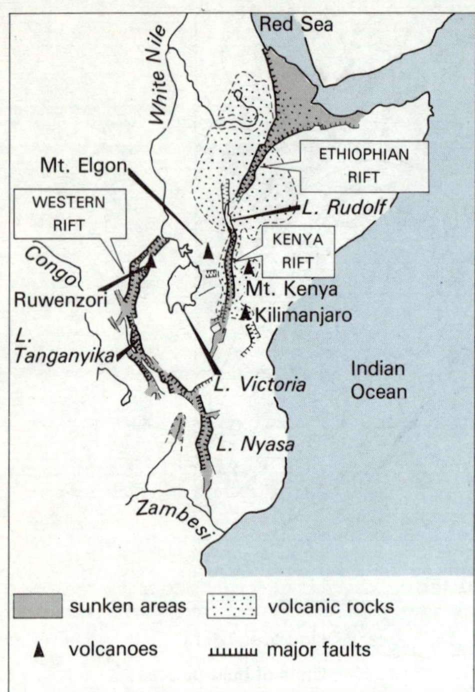

figure 8.13   *The East African Rift valley*

San Andreas Reservoir occupies a valley etched along part of the fault line. When earth movements cease such lines of weakness remain, to be exploited by later erosion. It was the gouging out of such fault lines by continental ice sheets which formed the rectilinear rock basins of the lake-studded Canadian Shield. Scotland's Great Glen, stretching from Fort William to Inverness, lies along a fault which occurred 350 000 000 years ago and which wrenched rocks about 90 km sideways!

The process of faulting can also produce scarp landforms. (Scarp is an abbreviation for escarpment, an abrupt change of relief.) Faulting with a vertical shifting of rocks can produce such a feature. The fault scarp rarely remains unaltered. Its face becomes fretted and dissected by weathering and erosion and its base often mantled with fans of debris.

One of the most spectacular landforms produced by faulting is a **rift valley**. Stretching more than 3000 km, the world's largest example is in East Africa (fig. 8.13). During Tertiary times a large crustal arch began to form. Tension or stretching on the flanks of this East African Dome produced two parallel series of down faults which produced a 30–50 km wide trough between them – the rift valley. The floors of the troughs were irregular so they now contain lakes such as Nyasa, Edward, Albert and the 1400 m deep Tanganyika in the western rift and Lake Rudolf in the eastern. The rift faults provided lines of weakness which allowed magma to escape to the surface and form the spreads of lava and the volcanic peaks of Kilimanjaro, Mt. Kenya, Mt. Elgon and Ruwenzori.

Closer to home, an older, smaller and much altered example of a rift valley is found in the Central Lowlands of Scotland. Today its most striking relief elements reflect volcanic intrusions and extrusions. These are relatively resistant compared to the surrounding carboniferous sediments and have produced the larger hill masses (e.g. the Ochils) as well as smaller remnants of volcanic plugs and dykes (e.g. Edinburg Castle rock).

Describe the formation of a rift valley.

Plate tectonics: the grand design

An issue in both first and last chapters is why the landforms produced by crustal processes are *where* they are, i.e. why there are fold mountains in western USA and a rift valley in Africa and volcanoes in Iceland. In other words, what has produced the *pattern of the major physical divisions of the earth's crust*?

A basic question

Earthquakes are produced by crustal motion so their distribution is a clue to the present pattern of earth movements. Volcanoes too are signs of crustal weakness. Figure 1.10 showed the world distribution of earthquakes and volcanoes. Notice two kinds of area: the fairly stable 'quiet' areas are known as **plates** and the active belts are the **plate margins**.

The plates, which extend to depths of up to 80 km and include areas of continental and oceanic crust, are in continual slow motion. Where they are moving apart new crust is forming. We studied this spreading zone situation in connection with the birth of Surtsey and the growth of Iceland as part of the Atlantic *mid-ocean ridge* (chapter 1). If new plate is being continually made, what happens to old plate? Its 'graveyard' is presumed to be the **ocean trenches** where one plate thrusts under another (see line of section, fig. 8.14). Where plates are in collision, massive fold mountain chains like the Himalayas and Alps are produced (a process reflected in fig. 8.7). The plates may also move past one another, a process which is occurring in California (fig. 8.10).

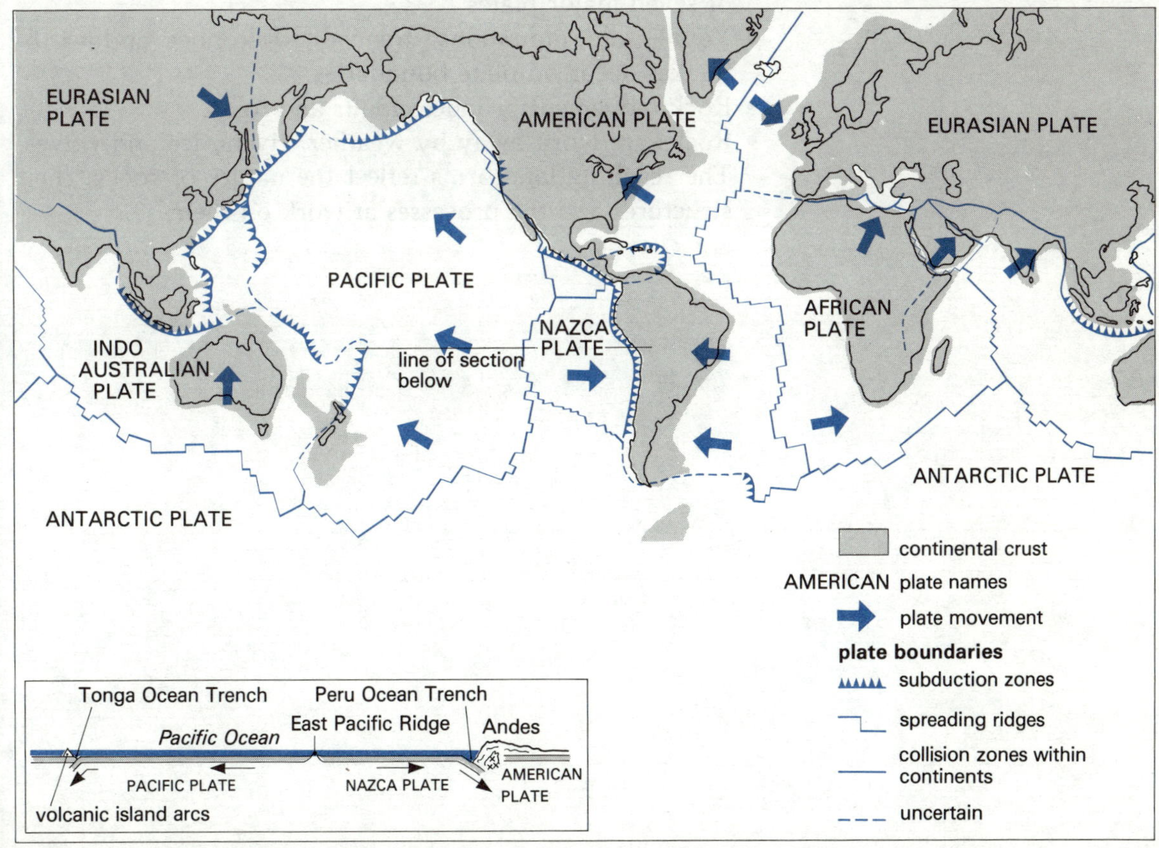

figure 8.14   Plate Tectonics: the grand design

Write a short essay describing Plate Tectonics Theory. Mention why it helps explain the world pattern of volcanoes, earthquakes, fold mountains, ocean trenches and mid-ocean ridges.

The names of the major plates are shown in fig. 8.14, together with arrows showing their present directions of movement. If you look at these arrows you can see that the Pacific is shrinking as the African, Antarctic and American plates grow at the rate of a few centimetres a year. If you look back to fig. 8.8 you can see that the present plate movements relate fairly closely to the pattern of young fold mountains and ocean trenches.

This view of the earth's surface as a series of plates is known as *Plate Tectonics Theory*. All the evidence for it is beyond our scope here, but it does reinforce at the largest scale the theme introduced at many places in the book – the earth is a restless place!

Summary

- The earth's crust, its outermost layer, is dynamic and capable of movement vertically and horizontally.
- The movement today is best explained by Plate Tectonics Theory. This views the earth (both land and sea) as a series of seven major plates.
- Young fold mountains, trenches, volcanoes and earthquakes occur on plate boundaries.
- Rock movement produces faults and folds.
- Rocks are worn away by weather, rivers, ice and waves. The resulting landforms reflect the nature of rocks, their structures and the processes at work on them.

Analysis of chapter content

Terminology and ideas	Skills and techniques	Location of detailed examples

1 volcanism an internal
    process
	following a chronology
	    from text, maps and
		Surtsey
		Surtsey

seasculpturing	block diagrams	Surtsey
plant colonisation	descriptive writing	Surtsey
weathering	imaginative writing	
volcanism, weathering	testing ideas	
erosion, transport	using a Landsat image	Iceland
deposition and landforms		
mid-ocean ridge	relating patterns	Atlantic
plate tectonics		world
rock cycle		world
water cycle		
2 water stores		
the drainage basin	map interpretation	
interception	modelling	
infiltration	scattergraphs	
overland flow	testing an idea	Somerset
throughflow	line graphs	
groundwater		

flow speeds in the basin
    system
	understanding a systems
	    diagram	

3 runoff variations	bar and line graphs	Devon
the storm hydrograph: water routing	hypothesis testing	Somerset
storm hydrographs	scattergraphs, prediction	Wales
dynamic basins	modelling, block diagrams and graphs	—

land use and river
    behaviour
	scientific method; map,
	    graph and systems
	    diagram interpretation
		Wye
		Harlow

water balances and river
    regimes
	bar graphs, pie graphs and
	    local area research
		Labrador
		Scotland

drought hazard aridity	scattergraphs, bar graph,
	    isoline maps and local
	    area research
		England
		—

4 landslide	geological cross sections	Italy
trigger, setting	writing a newspaper report	
rock classification		—
geological column		—
mechanical weathering	data interpretation	Kent
chemical weathering	short essay	Utah
mass movement	photo interpretation	Labrador
	classification	Alaska
		UK
karst	air photo interpretation	Florida
	examining theories	UK (Somerset)

The right-hand panel is a chart headed **Scale**, divided into **size** (small, medium, large) and **time** (days, months, years, centuries, millenia), with blue bars marking the relevant scale for each row.

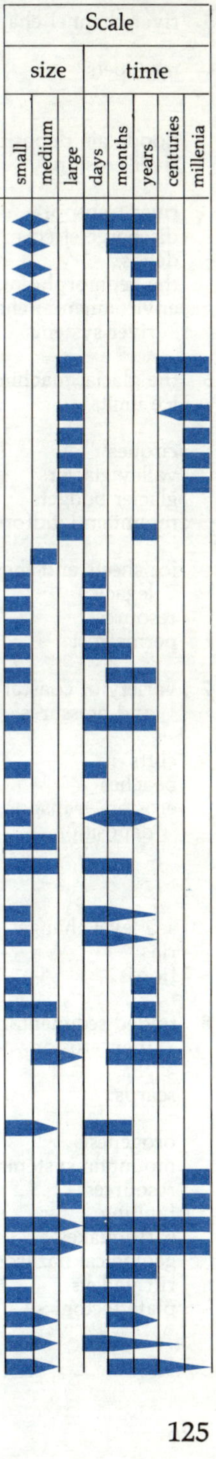

	Terminology and ideas	Skills and techniques	Location of detailed examples	small	medium	large	days	months	years	centuries	millenia
				size			**time**				
5	river channel changes	evidence interpretation and local area research	Mississippi	■	■		■	■	■		
	meanders	photo and diagram matching evidence interpretation	Wye		■			■	■		
		cross sections			■	■	■	■	■	■	
	floodplain deposition	bar graphs,	—		■		■	■	■	■	
	river erosion	describing relationships, photo interpretation	Devon	■				■			
	river transport	scattergraphs, map									
	discharge effects	interpretation, matching	Mackenzie		■						
	deltas	model to reality	Nile		■				■		
	the geomorphological cycle	modelling	—		◄						
	environmental change and river systems	line graphs	USA UK		■					■	
6	the glacial machine		Iceland	■							
	ice limits	map interpretation and reasoning	world					■			
	cirques	block diagrams		■	■						
	valley glaciers			■	■						
	glacier budgets	interpreting a report	Canada								
	mountain landform legacy	map and photo interpretation	Montana UK								
	ice sheets and their legacy	map description	N. England Labrador	■							
	resources	research	—								
	permafrost		—								
7	variety of coastal processes and pressures	text, map and photo synthesis local area research	Weymouth, Dorset	■			■				
	cliffs		UK	■					■		
	beaches		California	■				■			
	erosion, transport and deposition	discovering system interdependence and feedbacks	Dorset (Barton) Devon (Hallsands)								
		map text and photo synthesis	Yorkshire (Spurn)	■				■			
	sea level changes	interpreting evidence	Sussex (Cuckmere)	◄					■		
	rias	using a systems diagram									
	fjords	prediction									
8	folded sedimentaries and their erosion	cross sections	Hereford Weald Appalachians		■						■
	scarps	photo and map interpretation	UK	■							■
	orogenesis	modelling	Rockies		■						
	mountain systems		—								
	resources										
	faulting	interpreting data	San Andreas	■				■			
	earthquakes	reporting	fault system	◄			■			■	
	geological hazards		San Francisco		◄						
	rift valleys		E. Africa		■						■
	plate tectonics	synthesis	—		■						■

126

Glossary

ablation All water losses from a glacier, by ice melting and evaporation.

abrasion The erosive rubbing action of load, in a stream or glacier, or in the sea.

active layer Surface layer in the permafrost zone which thaws in summer.

alluvium Clay, silt and sand deposited across the valley floor by a river, especially after flood.

antecedent moisture Water in the soil before a storm.

anticline A folded arch of sedimentary rock.

arête The sharp ridge separating two cirques, often frost shattered.

arid Dry or desert areas, where evapotranspiration exceeds precipitation.

backwash Sea water returning down a beach after the breaking of a wave.

bar Low ridge of deposited silt or sand almost submerged by water – in rivers or sea.

baseflow That part of river flow provided by the groundwater stores. Unlike quickflows, relatively steady in the short term.

batholith A mass of igneous rock (e.g. granite) injected into older rocks, often exposed by later erosion.

bed load Sediment carried by streams along their beds in a rolling motion.

berm Ridge on a beach.

bioconstruction Building of landforms by organisms, e.g. silting encouraged by plants in delta backswamps.

bioerosion The erosion of rocks by organisms (plants and animals).

bluff A low cliff bordering a floodplain, left after meander migration.

braiding A stream channel pattern with islands, bars and shoals dividing channels.

breccia A sedimentary rock made up of cemented angular fragments.

budget Applied to glaciers, the idea of water 'income' (snow), 'expenditure' (ablation) and 'deposit' (the glacier's mass) for a year.

cementation The 'gluing together' of minerals in sedimentary rock.

chalk A soft, porous sedimentary rock composed of calcium carbonate.

cirque An armchair-shaped hollow produced by a small glacier.

compaction The 'pressing together' of minerals in the formation of sedimentary rocks.

cone Volcanic landform produced by falling tephra and/or outpourings of lava from a vent.

conglomerate A sedimentary rock made up of rounded fragments, like beach pebbles, cemented together.

corrasion Wearing away.

corrie *see* cirque.

crevasse A crack in the surface layers of a glacier.

crevasse splay A break in a levee.

crust Outermost layer of the Earth.

cumec Cubic metres per second, a measure of river discharge.

cycle of erosion The idea (W.M. Davis) that landscapes evolve through a sequence of 'stages' of youth, maturity and old age.

delta A composite landform, built of river sediment deposited by river distributaries in the sea and modified by sea action. Often triangular in shape.

dendritic A pattern formed by a stream and its tributaries, resembling the branches of a tree.

deposition The dropping of sediment by rivers, glaciers, wind and the sea.

dip The tilt (e.g. 15° NE) of strata in folded sedimentary rocks.

discharge A measure of stream flow, e.g. in cumecs.

distributaries Subsidiary channels carrying water away from the main channel in a delta.

diurnal Day/night.

dome Arched sedimentary rocks.

drainage basin The area drained by a stream and its tributaries, bounded by a divide or watershed.

drainage density Measure of the length of stream channels in an area (e.g. in km/km^2).

drumlins Glacially moulded deposits, streamlined in shape like upturned spoons; often occur in swarms.

dry valley Valley whose shape (section and profile) resembles a stream-carved valley but which has no surface water in it today. Common in limestone areas.

dyke An igneous intrusion where the magma has been forced up in narrow elongated fissures.

earthquake A violent shaking of the earth's crust produced by the sudden breaking of rocks.

emerged shoreline A shoreline which has apparently risen above the present sea level, either from land or sea level changes.

erg Sand 'sea' in the desert.

erosion Wearing away of rock.

esker A long ridge of sand and gravel, deposited by glacial meltwaters under the ice.

eustatic A change in sea level produced by a change in the volume of the oceans.

evapotranspiration The transfer of water to the atmosphere, by evaporation from water bodies and the soil and transpiration from plants.

fault Fracture in rock, offsets range from centimetres to kilometres.

felsenmeer A layer of angular rock fragments.

fetch The distance which wind may blow across open sea.

firn A 'halfway house' between light, fresh snow and denser glacier ice.

fissure An elongated crack through which magma reaches the surface.

fjord A 'drowned' glacial trough.

floodplain The flat land close to a river, built by river deposition and channel movements; liable to flooding.

freeze–thaw A mechanical weathering process where water in pores or cracks freezes and expands to fracture rock.

geosyncline A huge basin in which sediments accumulate in the early phases of orogenesis or mountain building.

glacier A mass of ice on the surface which shows signs of movement. Occur as small cirque glaciers, larger valley glaciers and as ice caps or sheets.

groyne A barrier built down a beach.

hanging valley In glaciated mountain areas the 'over-deepening' of glacial troughs leaves tributary valleys perched above the floor.

hazard An unusual, unpredictable natural event (e.g. flood, hurricane, earthquake) which harms people.

hogsback A sharp, pointed ridge.

hydraulic action The smoothing erosion of running water.

hydrograph A record of stream discharge.

hydrological cycle The movement of water (as gas, liquid or solid) between the various stores (air, rivers, lakes, soil, sea and glaciers).

hypothesis A statement of expected relationships.

ice cap An ice mass which blankets underlying hills and valleys.

igneous rocks Formed from magma cooling and solidifying on or beneath the surface.

infiltration The process of water seeping into the soil.

intensity Amount per unit of time, e.g. mm/hr for precipitation.

interception The catching or 'holding up' of precipitation by plant leaves and stems.

isostatic Vertical movement of the continents when their weight changes e.g. upwards when glaciers melt or mountains are eroded and downwards under the weight of a growing ice cap.

karst Limestone landscape, lacks surface drainage and has many solutional landforms.

lava Magma, molten rock, which has reached the surface.

levees Low embankments bordering a river, built by overbank deposition in times of flood.

limestone A sedimentary rock largely composed of calcium carbonate.

limestone pavement A flattish surface with little vegetation, etched by solution into a series of blocks and cracks.

load All the material (the visible clays, silts, sands and pebbles and the invisible dissolved minerals) carried by rivers.

long profile A trace of a river's height along its course.

longshore drifting The movement of beach materials along the shore because of oblique waves.

magma Molten rock.

marine erosion Wearing away by the sea.

mass movement A term including all the slow to fast, creeping, sliding, flowing movements of rock debris downslope.

meanders Series of sweeping curves in a river channel.

metamorphic rock Rocks altered by heat and pressure.

model A simplication of the real world. Can be diagrams, maps, equations etc. A tool to help us think.

monsoon Seasonal wind reversal.

moraine Rock debris carried by a glacier and deposited. **Lateral** moraines line the edge of moving glaciers. **Medial** moraines lie in the centre. **Terminal** moraines pile up at the glacier's snout to form a curving ridge across lower ground.

mudstone Sedimentary rock composed of very fine grained material.

ness A triangular lowland built of sand and shingle.

ocean trenches The deepest parts of the oceans. Belts where the ocean crust is drawn down (subducted) by plate motion at the 'collision' of oceanic and continental crust.

orogenesis The process of mountain building.

outwash Meltwaters pouring away from a glacier or ice sheet.

outwash deposits Clays, silts, sands and gravels carried by meltwaters and later deposited (e.g. as fans).

overland flow A quickflow, storm water unable to infiltrate into the ground which flows across the surface.

ox-bow lake A horseshoe-shaped lake formed when a river meander is cut off.

peneplain Almost a plain, i.e. with a few residual hills.

periglacial A zone where glacial meltwaters, snow and permafrost influence landforms.

permafrost Permanently frozen ground.

pillow lava Heaped lavas which have cooled and solidified beneath the sea.

plate A large relatively rigid section of the earth's crust that moves; may be continental, oceanic or a mixture.

plate margin Where moving plates meet. Can be convergent, spreading or sideways movements.

point bar On the inside edge of a meander, coarser deposits building a finger 'downstream'.

pool A deeper part of a river channel, e.g. on the outside bend of a meander.

pothole (a) A hole in a river's rock bed 'drilled' out by spinning pebbles.
(b) In limestone areas, name often used for vertical cave into which surface drainage disappears.

precipitation All the ways the atmosphere's water reaches the surface e.g. rain, snow, hail, drizzle etc.

pumice Molten rock foamed up by expanding gases in a volcanic eruption. Cools quickly to form a rock 'Aero bar'.

pyramidal peak A frost-fretted peak formed by the etching into a mountain of a series of cirques lying 'back to back'.

quarrying The removal of rock blocks by ice when a moving glacier plucks away bedrock frozen to the ice or by waves at the coast.

rejuvenation The 'renewed birth' or downcutting of streams, because of base level (i.e. sea) or climatic change.

resource Anything existing in nature used by people.

ria A 'drowned' river valley.

ribbon lake Elongated lake filling a depression in a glaciated valley floor. Produced by 'overdeepening' of the rock floor by moving ice and/or by damming behind a terminal moraine.

riffle Area of shallow water in a stream.

rift valley Valley bounded by faults, produced by large scale earth movements.

rill A miniature valley carved by rivulets on a slope.

roche moutonnée A product of glacial erosion: a rock mass whose 'up ice' side is often striated and 'down ice' side quarried.

runoff Within the drainage basin hydrological cycle, the water which leaves as streamflow. Often expressed in millimetres to help compare it to the precipitation input.

saltation A jumping motion of river load particles.

sandstone Sedimentary rock composed of cemented sand-sized particles.

scarp (a) An asymmetrical hill, a landform with a steep slope (scarp) and gentler slope (dip slope) on sedimentary rocks.
(b) A relatively straight steep slope or cliff produced by faulting.

scree A slope of small angular rock fragments produced by freeze-thaw weathering.

sedimentary rocks Rocks formed by the deposition of minerals in layers.

shale A sedimentary rock composed of fine particles.

shoal A shallow.

snowline The limit, on a mountain or glacier, above which winter's snow lies all summer.

soil creep Slow downslope movement of soil.

solifluction A sludging or slowish flowing of soil downslope, usually above permafrost.

solution load The removal of material from the landscape/drainage basin dissolved in the running water, e.g. calcium from limestone areas.

spall A layered, shelling form of rock breakdown.

spit A landform built by the sea (of sand, shingle or pebbles), attached to the land at one end, built by longshore drifting and enhanced by bioconstruction.

stack Isolated rock pillar on cliffed coast, previously part of the land.

store Where water or rock is left during its cycling, e.g. the groundwater store or the floodplain sediment store.

strata Layers in sedimentary rock.

striations Scratch marks on bedrock produced by the abrasion of load at the base of a glacier.

suspension load The material carried in suspension by a river.

swallow hole *see* pothole.

swash Water rushing up a beach after a wave breaks.

syncline A downfold in sedimentary rocks.

tephra Particles thrown out during a volcanic eruption.

terrace A flat belt of land within a river valley running parallel with the river but above it, the remnant of an earlier floodplain. Often produced by base level and climatic changes.

throughflow Water flowing through the soil towards the streams.

tombolo An island linked to the coast by a spit.

transfers Term applied to water movements between stores in the hydrological cycle.

transpiration The loss of water from plants.

trough A glaciated valley with steep walls, U-shaped cross profile and straight plan.

trough end Steep wall at the head of a glaciated valley where tributary cirques combined to form a powerful valley glacier.

truncated spurs Triangular shaped facets along the sides of a glaciated trough, marking the sites of pre-glacial interlocking spurs eroded by the valley glacier.

vent A single, central source of volcanic material.

volcanic activity The ways magma and gases escape from or move within the crust.

wadi A valley in arid areas only occasionally used by streams.

water table The upper surface of permanently saturated rock.

watershed Divide or limits of the drainage basin.

weathering Rock decay and destruction by weather, organisms and chemical processes.

Index